Warship 2019: Multi-Role Vessels

Bristol, United Kingdom
25-26 June 2019

ISBN: 978-1-5108-9542-3

Printed from e-media with permission by:

Curran Associates, Inc.
57 Morehouse Lane
Red Hook, NY 12571

Some format issues inherent in the e-media version may also appear in this print version.

The Royal Institution of Naval Architects
8-9 Northumberland Street
London, WC2N 5DA
United Kingdom

Phone: 020 7235 4622
Fax: 020 7259 5912

publications@rina.org.uk

Additional copies of this publication are available from:

Curran Associates, Inc.
57 Morehouse Lane
Red Hook, NY 12571 USA
Phone: 845-758-0400
Fax: 845-758-2633
Email: curran@proceedings.com
Web: www.proceedings.com

TABLE OF CONTENTS

ALLIED MARITIME FORCES TRANSFORMATION: TOWARDS A DEFENCE PLANNING PROCESS THAT INCLUDES ADAPTABLE WARSHIPS

Richard Logtmeijer, Ministry of Defence, NL
David Manley, Ministry of Defence, UK
Gregory Condon, Joseph Cole, Department of Defence, AUS
Jean-Denis Caron, Michèle Fee, DRDC, CA
Martin Tenbergen, Ministry of Defence, GE
Michael Rudius, MTG Marinetechnik GmbH, GE
Cdr Francesco Greco ITN, Ministry of Defence, IT
Siebe Otten, Ministry of Defence, NL
Julia Alvarez Mari, Naval Design Partnering Team, UK
Rod Pudduck, Jon Goodwin, Dickie Burston, Nate McKenzie, NIAG SG-236
Cdr Peter de Groot RNLN, NATO COE CSW

SUMMARY

Warships have long service lives. During the life of a warship the types of operations that will be assigned to the ship will change (this happened for example at the end of the Cold War), the technology behind the installed systems will advance (e.g., radar performance and miniaturisation) and new technologies will emerge. New technologies are likely to require changes in the way operations are presently conducted (e.g., off-board systems for conducting mine countermeasures operations) and can deliver new operational capabilities to the ship (e.g., directed-energy weapon systems). For these reasons, warships can only maintain maximum operational relevance through-life if their operational capabilities can be augmented and adapted to meet changing user requirements. NATO nations and partners, and also their peer competitors, are designing and building more adaptable warships. A common characteristic of these ships is that mission essential systems can be added to and removed from the ship in a relatively short time period. Warship roles can thus be reconfigured. The future of this trend is transforming the NATO defence planning process so that the future structure of the allied maritime forces will include an appropriate mix of adaptable warships and up-to-date mission packages that can respond to constantly changing operational tasking. The naval architect is already aware that traditional warship design features must be re-worked to accommodate modular, in addition to—or even to replace—organic systems. This paper considers the transformation from the engineering and management of mission packages, their deployment and integration into new warship designs towards a new maritime defence planning philosophy and process.

1. INTRODUCTION

At the RINA Warship 2018 conference a paper titled 'The NATO Drive to Mission Modularity' was presented [1]. This paper is also about NATO and the concept of mission modularity: the authors present their view on how mission modularity will transform the allied maritime forces. They all are or have been involved in NATO studies on mission modularity for the allied maritime forces.

The paper has the following structure. The paper starts with an explanation of what multi-role warships actually are according to the authors (Section 2). What is the difference between a multi-role warship, an adaptable warship, a flexible warship and a modular warship? Next, Sections 3 and 4 present the main reasons why nations and NATO are interested in developing and operating multi-role warships.

Before the authors can present their view on how mission modularity will transform the allied maritime forces, it is necessary to explain the NATO Defence Planning Process (Section 5). This process is executed every four years and delivers the range of forces that meets NATO's mission needs. The mission needs change as new threats and potential conflicts are identified. The capability building blocks used in the process include traditional single-mission and multi-mission warships. The transformation of the allied maritime forces will not happen without introducing new capability building blocks for multi-role warships.

An important topic of any RINA Warship conference is ship design (RINA: Royal Institution of Naval Architects). Section 6 of the paper describes the authors' view on the challenge of designing multi-role warships.

Section 7 gives an overview of NATO's recent mission modularity activities. Finally, in Section 8 the authors look into the future and present their view on the effects of mission modularity on the allied maritime forces. The expected effects are divided into three time periods: short-term effects (1 - 4 years from now), middle-term effects (5 - 8 years) and long-term effects (9+ years).

The paper ends with drawing conclusions from looking well into the future (Section 9).

2. MULTI-ROLE WARSHIPS

NATO nations and partners are designing and building multi-role warships. Multi-role warships are warships of which the combat capability can be disconnected from the ship and replaced with different combat capability. A multi-role warship can therefore conduct different types of operations, or in other words can have different warfighting roles. It however cannot host the combat capability required for all warfighting roles at the same time. This means that it may be necessary to reconfigure or re-role the combat capability of the warship before the next mission can be conducted.

A well-known example of a multi-role warship is the Littoral Combat Ship (Figures 1a and 1b) of the United States Navy (USN). Other excellent examples are the warships of the Royal Danish Navy (RDN) that implement the StanFlex concept (Figures 2a, 2b and 2c).

A warship that can host the combat capability required for different types of operations is called a multi-mission warship [2]. Multi-mission warships are considerably larger than multi-role warships.

Multi-role warships are described as being adaptable and flexible. The combat capability of a multi-role warship can be adapted relatively easy compared with a traditional single-mission or multi-mission warship. Relatively easy means that it will take less time and effort to update or replace the combat capability. The reason for this is that the combat capability of a traditional warship is integrated with the ship. Updating or replacing it is a complex on-board operation, whereas updating or replacing the combat capability of a multi-role warship starts with disconnecting it from the ship and moving it off-board, for example to a maintenance facility.

The cost of updating or replacing the combat capability of a traditional warship is high because it is a lengthy, complex on-board operation. As a result, the combat capability is not frequently updated or replaced, which may severely degrade the through-life operational effectiveness of the ship.

Because the combat capability of a multi-role warship can be adapted relatively easy, the ship offers flexibility in the types of operations it can conduct. A traditional warship has not been designed with adaptability in mind, and it is therefore much more difficult to modify its combat capability without a major update or refit. For example, there may not be enough free space to add a new combat system to the ship.

Multi-role warships are also described as being modular. The building blocks of multi-role warships are mission modules, potentially made up of a number of units. A unit is a mission system within the envelope of a standard-size 20-foot cargo container, also known as a Twenty-foot Equivalent Unit (TEU). There are exceptions; for example, a large boat or a helicopter does not fit within the envelope of a standard container. However, enclosing mission systems in standard shipping containers is an important advantage when moving these systems between the multi-role warship and locations on land such as mission module storage and maintenance facilities. Standard 20-foot cargo containers measure 20 feet in length, 8 feet in width, and 8.5 feet in height. A standardized form factor is designed to allow them to be loaded on ships, trucks, railroad cars and heavy lift aircraft [2].

A group of mission modules that can provide a distinct combat capability is called a mission package. For example, an Anti-Submarine Warfare (ASW) mission package contains the mission modules that are required for ASW operations. One of the mission modules may be a containerised sensor that can be towed by the ship when it is deployed at sea.

The concept of disconnecting combat capability from the ship and replacing it with different combat capability is called maritime mission modularity. In this paper mission modularity means maritime mission modularity, and multi-role warship, adaptable warship, flexible warship and modular warship all have the same meaning. Mission modularity does not relate to warship modular construction.

3. NATIONS' INTEREST IN MULTI-ROLE WARSHIPS

NATO nations and partners are designing and building multi-role warships. There are four reasons why nations nowadays want multi-role warships instead of single-mission or multi-mission warships: current capability requirement, warship acquisition cost, technological advancements and uncertainty of the capability requirement.

3.1 CURRENT CAPABILITY REQUIREMENT

The current threat situation demands for high-end warfighting capabilities. High-end warfighting capabilities include capabilities for ASW, Anti-Surface Warfare (ASUW) and Anti-Air Warfare (AAW). These capabilities are called primary capabilities. A nation also needs capabilities for other types of operations, for example Maritime Interdiction Operations (MIO), Non-combatant Evacuation Operations (NEO) and Humanitarian Assistance and Disaster Relief (HADR) operations. These capabilities are called secondary capabilities. A cost-effective way of meeting the current capability requirement is for a nation to design and build capable and multi-role warships. The multi-role part of these ships may be limited to secondary capabilities. For example, a warship may be configured for NEO or for HADR operations. Although more difficult to implement, the multi-role part of these capable and multi-role warships may also include primary capabilities. For example, a warship may be configured for ASUW + AAW or for ASUW + ASW. The USN Littoral Combat Ship (Figures 1a and 1b) is an example of a multi-role warship of which the primary capabilities can be reconfigured.

3.2 WARSHIP ACQUISITION COST

By developing and operating multi-role warships, it may be possible to reduce the number of ships required to meet mission, and therefore warship acquisition cost will also reduce. Multi-role warships can be reconfigured for different types of operations. This means that not all combat capability required to meet mission needs has to be on board of the warships; part of it can be stored on land.

3.3 TECHNOLOGICAL ADVANCEMENTS

The design of a new warship starts with identifying a spectrum of missions. The ship must be capable of successfully completing the full spectrum of missions, so the spectrum of missions is translated into required combat capability, which is subsequently translated into mission systems. It takes many years before the new warship is ready for sea for war. And the service life of the new ship may well be over thirty years. This means that after some time the technologies of the mission systems may have advanced. Without updating or refreshing those systems, there is a risk that the ship's combat capability is no longer effective against new threats. The combat capability of a multi-role warship can be adapted relatively easy compared with a traditional single-mission or multi-mission warship. This means that it is relatively easy to replace outdated mission systems with new mission systems that make use of technological advancements.

3.4 UNCERTAINTY OF THE CAPABILITY REQUIREMENT

As mentioned under 'Technological Advancements', it takes many years before a new warship is ready for sea for war. And the service life of the new ship may well be over thirty years. After some time the spectrum of missions that will be assigned to the ship may have to change because of new threats and new disruptive technologies. This means that the required combat capability and mission systems also have to change. The combat capability of a multi-role warship can be adapted relatively easy. New mission systems that are capable of countering the new threats and new disruptive technologies can be designed and built on land, and once finished installed on board without complex integration.

4. NATO'S INTEREST IN MULTI-ROLE WARSHIPS

Just like individual nations, NATO periodically identifies threats and potential conflicts, and defines a minimum capability requirement. The 29 member nations collectively must be capable of meeting the minimum capability requirement. It is in NATO's interest to have a range of forces that meets the minimum capability requirement in the most cost-effective way. Therefore the four reasons why nations nowadays want multi-role warships (mentioned in the previous Section) are also important to NATO. There are two other reasons why NATO is interested in this trend: interoperability and distribution of combat capability.

4.1 INTEROPERABILITY

Multi-role warships are warships of which combat capability can be disconnected from the ship and replaced with different combat capability. Interoperability means that a multi-role warship of nation A can be equipped with a mission package from nation B. To make this possible, the warships and the mission packages must comply with NATO standards. These standards specify for example the connections between warships and mission modules for exchanging electricity, cooling water and computer data. A nation can contribute a mission package to a NATO operation provided the mission package meets the NATO standards. Multi-role warships and interoperability lower the threshold for nations to contribute to NATO operations.

4.2 DISTRIBUTION OF COMBAT CAPABILITY

The distribution of the required combat capability between the NATO members can be controlled at a finer level. NATO nations and partners are developing and operating multi-role warships, and consequently the range of forces will include multi-role warships and mission packages. The number of multi-role warships and the number of mission packages are disconnected: there can be more mission packages than warships, for example. This means that at the NATO level there can be finer control over the required number of mission packages for a specific combat capability. Important is that a nation can make a mission package available to NATO, instead of a warship. For example, if each NATO member invests in Mine Countermeasures (MCM) vessels, from a NATO point of view there could be a surplus of that specific capability. Now the NATO members collectively can own exactly the number of mission packages for MCM operations as specified by NATO's minimum capability requirement. It is conceivable that in the future NATO owns a pool of mission packages for various types of operations, and NATO members share these for training, exercises and future conflicts.

5. THE NATO DEFENCE PLANNING PROCESS

It is in NATO's interest to have a range of forces that meets the minimum capability requirement in the most effective way. The process to obtain this result is called the NATO Defence Planning Process (NDPP). An overview of NDPP is necessary in order to describe the effect that multi-role warships have on the defence planning process. First, there is a difference between defence planning and operations planning [3]:

- Defence Planning is about trying to ensure that, if needed, the Alliance and its operations planners will have the required Capabilities available to meet its objectives in an uncertain future (medium-term to long-term).
- Operations Planning is about planning the use of existing forces and capabilities for a known and existing or imminent contingency (short-term) – Force Generation.

The NDPP consists of five steps [4]:

Step 1 Establish Political Guidance

First, NATO identifies threats and potential conflicts, and establishes political guidance for the defence planning effort.

Step 2 Determine Requirements

Secondly, NATO defines the Minimum Capability Requirement (MCR). The MCR is compared with the capabilities available to NATO and important capability shortfalls are identified.

Step 3 Apportion Requirements and Set Targets

Next, the MCR is distributed between the NATO members in the form of target packages. There is emphasis on the important capability shortfalls.

Step 4 Facilitate Implementation

In this step NATO assists members in implementing the target packages, for example by conducting internal research and development and providing NATO standards.

Step 5 Review Results

In this step NATO reviews the ability of the allied maritime forces to meet NATO standards and mission needs.

Steps 1, 2 and 3 are executed in sequence every four years. Step 5 is conducted every two years. Efforts to facilitate implementation (Step 4) are continuous and conducted in parallel with the other steps (Figure 3).

An important tool for executing the NDPP is the NATO capability codes and capability statements (CC & CS) document [5]. The CC & CS specify reference capabilities such as Warship Capable (NWC) and Anti-Air Warfare (NAAW). The capabilities of existing warships can be compared with the reference capabilities and the differences can be expressed numerically using equivalency factors. For example, the AAW capability of an existing warship may be 1.2 times the reference capability NAAW.

The building blocks for executing the NDPP are the capability codes. These are used in steps 3 and 5 to translate between the range of real forces available to NATO and the range of reference forces required to meet NATO standards and mission needs.

At the moment there are no CC & CS for multi-role warships and mission packages (except for a harbour protection mission package). This means that at the moment the NDPP cannot deliver a range of forces that includes multi-role warships and mission packages.

6. SHIP DESIGN

There is an obvious solution to this problem: create capability codes and capability statements for multi-role warships and mission packages. To do this, a set of reference multi-role warships and mission packages have to be designed. The reference multi-role warships could be of different sizes. The reference mission packages are for different types of operations. The challenge is to design a set of reference multi-role warships and mission packages that enables the NDPP to deliver a range of forces that meets the minimum capability requirement in the most cost-effective way.

7. NATO'S RECENT MISSION MODULARITY ACTIVITIES

NATO's recent mission modularity activities can be divided into three parts:

- work done by specialist teams of the Ship Design Capability Group (SDCG);
- internal research and development conducted by subgroups of the NATO Industrial Advisory Group (NIAG);
- work done by other NATO bodies.

7.1 SHIP DESIGN CAPABILITY GROUP

There are three specialist teams of the SDCG actively working on mission modularity: the specialist team on mission modularity, the specialist team on total ship systems engineering and the specialist team on ship costing (ST-MM, ST-TSSE and ST-SC respectively).

- ST-MM coordinates the activities of the specialist teams that contribute to Smart Defence initiative SD 1.41 on mission modularity. ST-MM also writes NATO standards on mission modularity, using the contributions of the specialist teams and NIAG subgroups (discussed below). There are now two NATO standards available:
    - Standardization Recommendation (STANREC) 4806 covering Allied Naval Engineering Publication (ANEP) 91 on standard interfaces for mission modules [6, 7];
    - Standardization Agreement (STANAG) 4830 covering ANEP 99 on design and interface standards for containerised mission modules.

  The standards can be downloaded from the website of the NATO Standardization Office.

- ST-TSSE and ST-SC conduct a cost-benefit analysis (CBA) of mission modularity for allied maritime operations. The CBA compares the cost and operational effectiveness of two task forces, one composed of traditional warships and one composed of modular warships. Both task forces have to conduct the same set of missions. To conduct the CBA, computer modelling and simulation is used (Figure 4). Two models have been developed: one for simulating the operational effectiveness of a given task force, and one for estimating the acquisition, operations and support costs of that task force. The expertise of naval operators has been used to construct the set of missions. Also, multi-role warships of different sizes and mission packages for different types of operations have been designed. Initial results show that for a given set of missions a relatively small modular task force will be more effective than an equally small traditional task force. There are no results yet on the cost difference between the modular task force and the traditional task force. Final results are expected by the end of this year.

7.2 NATO INDUSTRIAL ADVISORY GROUP

Since 2010 NIAG subgroups have studied various aspects of mission modularity:

- SG-150: 'Joint Support Ships / LPD Type Ships - Potential for Interoperability of Deployed Systems. A Modularity Assessment' [8];
- SG-168: 'Development of Modular Mission Packages for Joint Support Ship/LPD Type Ships' [9];
- SG-186: 'Modularity Demonstration and Validation' [10];
- SG-213: 'Mission Modularity Integrated Logistics Support' [11];
- SG-236: 'Scenario Development for Mission Modularity Force Architecture for Naval Platforms'.

SG-150

Initial interest focused on the example of the USN Littoral Combat Ship that introduced the concept of 'mission swapping' using bespoke combat systems embarked in ruggedized containers (ISO dimensions for ease of transportation and handling) designed by Northrop Grumman.

The LCS can embark a choice of containerised modules appropriate to the mission need for MCM, Surface Warfare (SUW) and ASW scenarios (Figure 5). Other mission packages not embarked in containers include helicopters and a marine detachment for counter-piracy/anti-terrorist operations.

The 2010 Haiti earthquake had demonstrated the difficulties of disembarking heavy equipment from ships when the port facilities are disabled, so the first NIAG study considered Joint Support Ship / Landing Platform Dock (LPD) type vessels since these offered the most suitable space and handling capabilities with Landing Craft Utility (LCU) and causeways for landing on a beach.

The ubiquity and cheapness of the ISO container suggested wide application for transportation and outfitting to form a 'module' and for the SG-150 study the standard TEU was considered for civil-military co-operation (CIMIC) operations, viz., HADR, Harbour Protection (HP) and Counter Piracy (CP). The ease of pre-loading with all types of equipment such as communications equipment, field hospitals, motor generators, reverse osmosis plants, to blankets, dried and canned food, and even a brig (detention unit/cells), demonstrated a means of achieving improved readiness and prompt delivery of a required capability to theatre by naval vessels equipped to carry mission modules.

SG-168

It was quickly realised that space, weight and handling arrangements were not the only components to be considered when installing a module in a ship. Two other major factors were introduced for the next study SG-168: interoperability (a NATO priority) and the utility of modules for warfighting scenarios. Search and rescue in extreme conditions was an additional scenario considered for the potential to augment smaller vessels with special equipment.

The overall objective of the SG-168 study was to develop a high-level Interface Control Document (ICD) to identify all ship-module interfaces. The high-level ICD aims to serve as the basis for a NATO Standard for Mission Modularity to allow nations to acquire interoperability elements tailored to specific national missions that can be installed as plug-and-play.

Figure 6 illustrates the Prime Item Development Specifications (PIDS) and ICDs that must subsequently be developed for a single containerised power generation module. When a modular fit is selected for a given mission scenario that may comprise a number of associated modules and modular units, the ICD becomes more complicated.

The requirement for interoperability is a utopian concept which obviously needs standardisation of interfaces either at fleet level or by all participating navies. It also raises difficult questions about ownership of the modules, logistics, maintenance and regeneration. There are several examples where standardisation of interfaces within NATO navies has already been achieved, notably refuelling at sea and helicopter deck landing arrangements. The NATO Submarine Rescue System (NSRS) is a good example of a shared modular capability.

SG-186

The primary requirement of this study was to develop interface design criteria for the development of a NATO standard for the installation and operation of mission packages in naval surface combatants, for the enhancement of mission capability and interoperability with other Alliance units. Secondly, the study was required to include illustrative concepts for shock attenuation of mission modules in both natural and warfighting environments, and associated interfaces and processes to support interoperability. Thirdly, it was required to produce an ICD based on selected concept designs, for a future full-scale shock test and verification trial and simulation, which could make use of synthetic environment techniques.

To capture the required interfaces a table was produced of all the potential mission modules and the interfaces to those modules. Research then determined the existing standards governing those interfaces (for compatibility) and a template for the ICD was produced. It is anticipated that the ICD will expand over time to incorporate new interface types and change when legacy interfaces become obsolete. The ICD template will act as a prompt to ensure that the data to define new interfaces remains consistent and to the required level of detail. Not all aspects in the template will be applicable to a given interface so it is more of an aide memoire. Like the ICD, it may be updated or expanded in the future.

The ICD from SG-168 was then taken and updated to follow the ICD template. Without detailed knowledge of modules which presently do not exist, an ICD was produced which should meet the majority of interface requirements. Analysis of the interfaces was conducted highlighting the universal interfaces (as covered by the ICD) and variants or interface types which are liable to be needed for a small range of modules. Recommendations for the additional interfaces were included in the ICD.

SG-186 built on the work in SG-168, which proposed an outline cost-benefit methodology based on an established Life Cycle Costing (LCC) model. This was considered a viable approach in the absence of comprehensive ship/module cost data. The problem of access to cost data remained in SG-186 which therefore sought to develop a methodology which was based on relative costs, to aid selection between conventional and modular ships.

A viable cost methodology for identifying acquisition and support cost drivers and comparing the cost of modular options was identified that can be applied from the earliest stages of the acquisition cycle and where there is limited or no cost data available. A qualitative costing methodology may be adapted for a range of missions and number of ships considered in any scenario. A main conclusion of the modularity costing exercise is unsurprisingly, the more system complexity, the higher the cost.

Costing the ownership of a modular warship is highly dependent on its utility, and this has caused some criticisms of the LCS. However, the question is complicated because, at least from a NATO perspective the comparison should be between a conventional and modular fleet rather than by analysis of a single vessel or class.

The UK Ministry of Defence sponsored and funded the development and demonstration of a candidate shock protection system for mission modules based on ISO containers. The shock protection system was designed for the ISO container to be fixed to ten shock mounts of an innovative design – known colloquially as 'pumpkin mounts'.

SG-213

In order to assist in the evaluation and full understanding of the merits of modularisation, more detailed work was needed. Hence SG-213 considered the whole gamut of logistics. A logistic support analysis strategy was proposed and a Generic Mission Module Logistics Process (GMMLP) developed, to address ownership and authority for the sharing of modules and the use of host ships, storage locations and port access for installing/changing modules.

Notwithstanding the need for compliance with established NATO doctrine and existing Integrated Logistics Support (ILS) documentation and the complexity of a yet to be implemented Alliance policy of holding mission modules at pre-determined bases and their deployment to prepared modular ships, the proposed GMMLP showed how such a system might be managed in the future.

Building on earlier NIAG studies SG-213 considered the basic logistics of transporting modules from base depot to ship, the effects of positioning on the ship design and key cost drivers. Mission module logistics can be considered in three phases (Figure 7):

- Setup: planning and preparation of the mission module;
- Operational usage: delivery of the mission module, installation and use;
- Return to storage: return of the mission module to storage/regeneration.

SG-236

Commenced in February 2019 this study pulls together all the previous NIAG mission modularity studies. This on-going study aims to develop and exercise common naval scenarios for CIMIC and warfare NATO missions to aid in the determination of the utility and force architecture construct of ship-based mission modules by means of an available computer model. An assessment will be made of available naval fleet computer models on their ability to measure sensitivities of their calculated cost and benefit analyses of mission modularity in order to identify a suitable model for the study. It will also evaluate whether the use of mission modules changes or impacts on the number of naval assets required and their inherent performances (i.e. trans-warship requirements).

There is no straightforward answer to the cost-benefit comparison between conventional and modular ships. The permutation of ship sizes and numbers, the positioning of mission modules ashore, how many and what types of modules should be built and who owns them are some of the presently unanswered questions that complicate matters. With the exception of historical data, obtaining reliable cost data relevant to new and emerging technologies will always be difficult. Predicting future costs for yet to be developed new types of ship, modular systems, fitting/exchanging modules and associated new strategies for ownership, basing and overall management (either by individual nations or by the Alliance) poses and almost intractable problem.

The metrics for evaluating cost-benefit of mission modularity must include the uncertainty that is being addressed. It is anticipated that any candidate assessment tool will generate these metrics on a Monte Carlo or similar type simulation where the uncertainty is modelled as a probability density function. This type of modelling needs to include a dynamic model of the management action (or choice model) which would determine how the modularity would be exploited over the ship's service life.

Finally, the study will summarise the key conclusions of all the NIAG mission modularity studies to date to provide a comprehensive overview of the merits and costs of introducing, managing and operating mission modules and modular capable naval platforms.

7.3 OTHER NATO BODIES

Other NATO bodies are also studying mission modularity, for example:

- There is a specialist team on mission modularity for MCM (ST-MMMCM). Many nations are switching from minehunters and minesweepers (traditional single-mission warships) to MCM vessels from which off-board MCM systems are operated.
- The NATO Centre Of Excellence for operations in Confined and Shallow Waters (COE CSW) is leading a NATO Smart Defence study called 'Flexible Interoperable Toolbox meeting Future Operational Requirements for Confined and Shallow Waters' (FIT FOR CSW). The objective of this study is 'To foster synergy in developing, procuring and operating a set of compatible components, thus enabling nations to maintain a broad(er) set of affordable and effective capabilities for maritime Combined and potentially also Joint operations, especially in the particular challenging environment of CSW.'

8. EFFECTS OF MISSION MODULARITY

In this Section the authors look into the future and present their view on the effects of mission modularity on the allied maritime forces. The expected effects are divided into three time periods: short-term effects (1 - 4 years from now), middle-term effects (5 - 8 years from now) and long-term effects (9+ years from now).

8.1 SHORT-TERM EFFECTS

According to the authors, the short-term effects (1 - 4 years from now) of mission modularity on the allied maritime forces are the following:

- New multi-role warships and mission packages will appear, for example the Royal Navy Type 26 (Figure 8a) and the Italian Navy Pattugliatori Polivalenti d'Altura (PPA) (Figure 9a). The first of class of the PPA was launched June 15th. These ships are equipped with mission bays, cranes for embarking and disembarking mission modules, and transport systems for moving the mission modules around the mission bay [12]. A mission bay is a space of the ship designed for storing and operating mission modules (Figures 8b and 9b). It has interface panels that are used for connecting the mission modules with the ship's systems, such as the electrical system, cooling water system and computer systems.
- However, the interoperability of the new multi-role warships and mission packages will be limited. The programmes for designing and building these ships have started before the publication of the NATO mission modularity standards.
- New editions of the NATO mission modularity standards will appear (STANREC 4806 and ANEP 91, STANAG 4830 and ANEP 99).
- The results of NATO's internal research and development activities will be published. For example, ST-TSSE and ST-SC and also NIAG SG-236 will publish the results of their cost-benefit analyses of mission modularity for allied maritime operations.
- The benefits of mission modularity for NATO will be demonstrated through participation in NATO exercises.
- New Doctrine, Organization, Training, Material, Leadership, Personnel, Facilities and Interoperability (DOTMLPFI, a NATO acronym) will appear. Importantly, new multi-role warship designs alone will not deliver a new capability to the allied maritime forces. It is equally important to develop new doctrine for operating the multi-role warships. And to develop a new way of managing personnel, especially the crews assigned to mission packages. Relevant factors for delivering a new capability are DOTMLPFI. The last factor, interoperability, is particularly important to NATO. Interoperability can be achieved through

standardisation on all of the other DOTMLPFI factors.

- A NATO standard for early stage warship design will be developed. The standard will assist NATO members in designing cost-effective warships. It will also assist NATO in designing a cost-effective range of forces that meet mission needs. This will be achieved by using the standard for designing reference types of multi-role warships and mission packages.

8.2 MIDDLE-TERM EFFECTS

The expected middle-term effects (5 - 8 years from now) of mission modularity on the allied maritime forces are the following:

- The authors expect that the trend of designing and building multi-role warships will increase, because of the reasons described in Sections 3 and 4. More multi-role warships and mission packages will appear, and traditional warships will disappear.
- The new multi-role warships and mission packages will be interoperable as NATO members will implement the NATO mission modularity standards.
- The NATO mission modularity standards will become mature and require only minor revisions.
- The effects of mission modularity on the cost-effectiveness of the partially transformed allied maritime forces will become clear. It is expected that these effects are positive; it will be interesting to see whether these effects are the same as predicted by the cost-benefit analyses conducted by ST-TSSE, ST-SC and NIAG SG-236.
- NATO will create an inventory of mission packages for various types of operations. NATO members can share these mission packages for training, exercises and future conflicts.
- New capability codes will be derived from the reference types of multi-role warships and mission packages, and these new capability codes will be used in the next iterations of the NDPP.

8.3 LONG-TERM EFFECTS

Finally, the expected long-term effects (9+ years from now) of mission modularity on the allied maritime forces are the following:

- The allied maritime forces have been transformed into a maritime force that is modular and more cost-effective.
- The NATO members will share an inventory of mission packages for training, exercises and future conflicts.
- The NDPP will be capable of delivering a cost-effective range of forces that meets mission needs. New mission packages can be developed in less time than new warships. So because of the introduction of the new capability building blocks for multi-role warships and mission packages, NATO will be capable of responding quicker to changes in mission needs as new threats and potential conflicts are identified.

9. CONCLUSIONS

- There are four reasons why nations nowadays want multi-role warships instead of single-mission or multi-mission warships: current capability requirement, warship acquisition cost, technological advancements and uncertainty of the capability requirement.
- There are two other reasons why NATO is interested in the trend of developing and operating multi-role warships: interoperability and distribution of combat capability.
- It is in NATO's interest to have a range of forces that meets the minimum capability requirement in the most cost-effective way. The process to obtain this result is called the NATO Defence Planning Process (NDPP). The building blocks for executing the NDPP are the NATO capability codes and capability statements (CC & CS). These are used in steps 3 and 5 of the NDPP to translate between the range of real forces available to NATO and the range of reference forces required to meet NATO standards and mission needs. At the moment there are no CC & CS for multi-role warships and mission packages. This means that at the moment the NDPP cannot deliver a range of forces that includes multi-role warships and mission packages.
- The challenge is to design a set of reference multi-role warships and mission packages that enables the NDPP to deliver a range of forces that meets the minimum capability requirement in the most cost-effective way.

- The expected short-term effects (1 - 4 years from now) of mission modularity on the allied maritime forces are: interoperability will be limited; the benefits of mission modularity for NATO will be demonstrated through participation in NATO exercises; new DOTMLPFI will appear; reference types of multi-role warships and mission packages will be designed.

- The expected middle-term effects (5 - 8 years from now) are: new multi-role warships and mission packages will be interoperable; the positive and negative effects of mission modularity will become clear; NATO will create an inventory of mission packages for various types of operations; new capability codes for multi-role warships and mission packages will be used in the next iterations of the NDPP.

- The expected long-term effects (9+ years from now) are: the allied maritime forces will have been transformed into a maritime force that is modular and more cost-effective; the NATO members will share an inventory of mission packages for training, exercises and future conflicts; the NDPP will be capable of delivering a cost-effective range of forces that meets mission needs; NATO will be capable of responding quicker to changes in mission needs as new threats and potential conflicts are identified.

10. DISCLAIMER

The opinions expressed in this paper are solely those of the authors; they do not necessarily represent the opinions of NATO or the organisations for which the authors work.

11. ACKNOWLEDGEMENTS

The authors would like to thank Jennifer Lin and Scott Clarke, NAVSEA, US, for their support.

12. REFERENCES

1. MANLEY, D., 'The NATO Drive to Mission Modularity', *Transactions of Warship 2018: Procurement of Future Surface Vessels*, 2018.

2. RAND National Defense Research Institute, 'Littoral Combat Ships: Relating Performance to Mission Package Inventories, Homeports, and Installation Sites', 2007.

3. NATO Allied Command Transformation, 'The NATO Defence Planning Process (NDPP)', NATO UNCLASSIFIED, 2019.

4. NATO, 'NATO Defence Planning Process', https://www.nato.int/cps/en/natolive/topics_49202.htm, Accessed on 13 June 2019.

5. NATO Committee for Standardization, 'Capability Codes and Capability Statements', NATO UNCLASSIFIED Releasable to EU, PfP, MD, ICI, Afghanistan, Australia, Iraq, Japan, The Republic of Korea, Mongolia, New Zealand, and Pakistan, 2016.

6. NATO Standardization Office, 'STANREC 4806 Standard Interfaces for Mission Modules', NATO UNCLASSIFIED Releasable to FIN, SWE, UAE and PAG countries, 2017.

7. NATO Standardization Office, 'ANEP-91 Standard Interfaces for Mission Modules', NATO/PfP UNCLASSIFIED, 2017.

8. NIAG SG-150, 'Joint Support Ships / LPD Type Ships - Potential for Interoperability of Deployed Systems. A Modularity Assessment', NATO UNCLASSIFIED, 2011.

9. NIAG SG-168, 'Development of Modular Mission Packages for Joint Support Ship/LPD Type Ships', NATO UNCLASSIFIED Releasable to PFP, 2014.

10. NIAG SG-186, 'Modularity Demonstration and Validation', NATO UNCLASSIFIED, 2016.

11. NIAG SG-213, 'Mission Modularity Integrated Logistics Support', NATO UNCLASSIFIED Releasable to PFP, 2017.

12. GRECO, F. and SERPAGLI, S., 'Definition and Development of the Modularity Features for the Italian Navy Multirole Patrol Vessel Mission Bays', *Proc. of the Technology and Science for the Ships of the Future Conference*, 2018.

13. AUTHORS' BIOGRAPHY

Richard Logtmeijer holds the current position of Senior Staff Member Life Cycle Modelling at the Defence Materiel Organisation (DMO) of the NL Ministry of Defence. He is responsible for analysing through modelling and simulation the life cycle effectiveness and cost of new warship concept designs. He is also chairman of the NATO specialist team on total ship systems engineering (ST-TSSE). His previous experience includes three years serving as a naval officer and over twenty years working for the Royal Netherland Navy.

David Manley holds the current position of Naval Authority Group Deputy Head, Ships Division in the UK Ministry of Defence. He is responsible for the development and management of the MOD's maritime combat safety programme as well as the Royal Navy's Survivability Strategy. His previous experience includes sixteen years working in maritime survivability and over twenty-five years working in the MOD. He is a DE&S Defence Engineering Senior Fellow, a Fellow of the Royal Institution of Naval Architects and a member of the Royal Corps of Naval Constructors.

Figure 1a: USN Littoral Combat Ship: USS *Independence*

Figure 1b: USN Littoral Combat Ship: USS *Freedom*

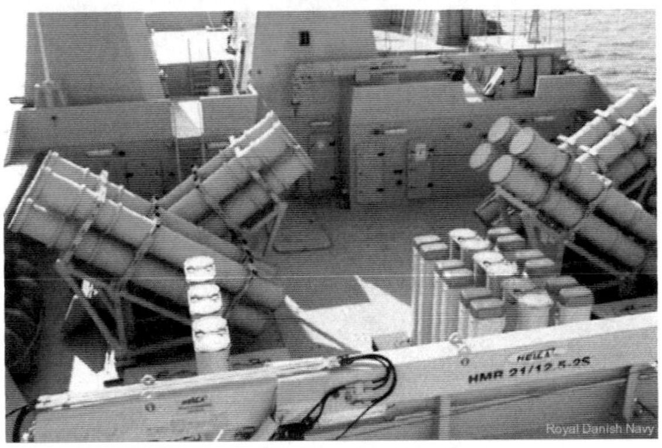

Figure 2a: RDN StanFlex warships: mission module

Figure 2b: RDN StanFlex warships: Absalon class support ship

Figure 2c: RDN StanFlex warships: Iver Huitfeldt class frigate

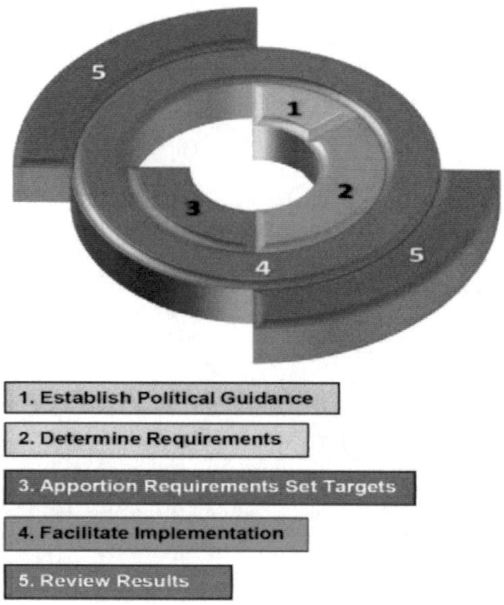

Figure 3: NATO Defence Planning Process

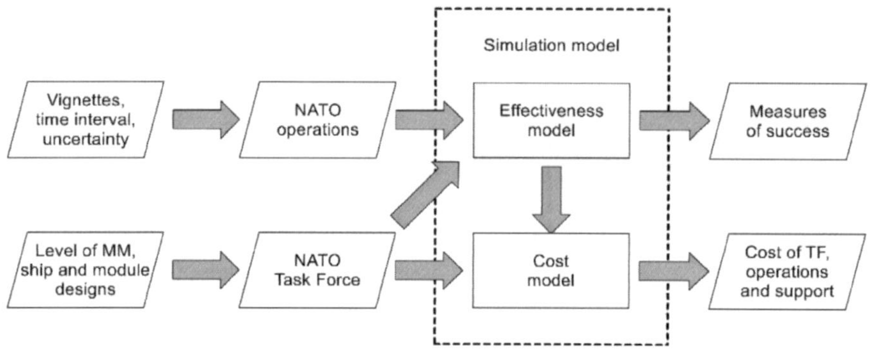

Figure 4: Computer modelling and simulation for conducting the CBA

Figure 5: Containerised modules for the USN Littoral Combat Ship

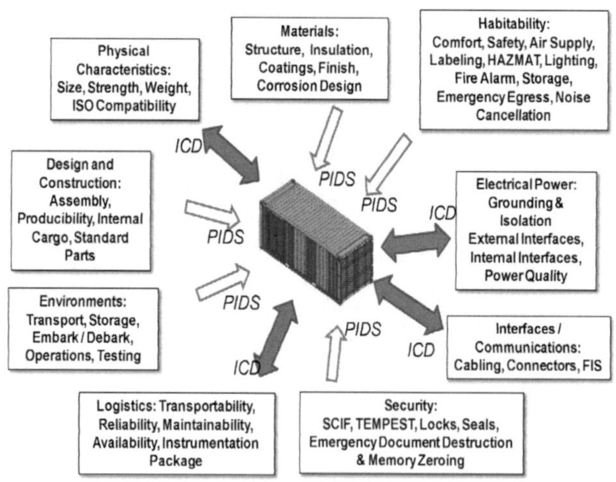

Figure 6: Single containerised power generation module

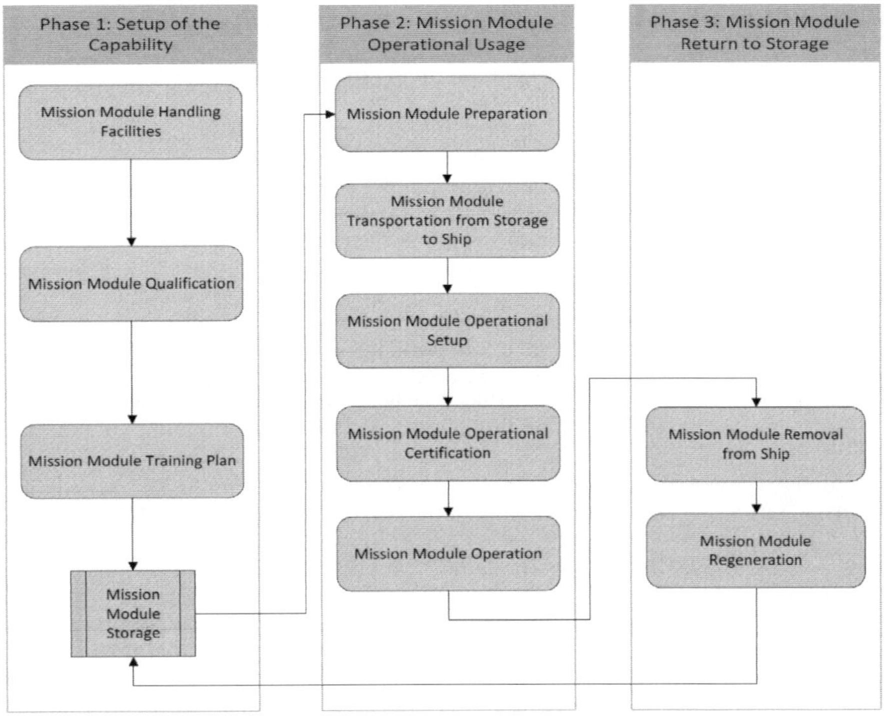

Figure 7: Generic Mission Module Logistics Proces

Figure 8a: RN Type 26

Figure 8b: RN Type 26: mission bay

Figure 9a: ITN Pattugliatori Polivalenti d'Altura *Thaon di Revel*

Figure 9b: ITN Pattugliatori Polivalenti d'Altura: aft mission bay

MULTI-ROLE VESSELS: DESIGN, MANNING & COST DRIVERS

D R Smith, MOD Naval Design Partnering, Bristol, UK

SUMMARY

Informed by the Naval Design Partnering involvement in concept design for future UK warships and auxiliary vessels, this paper comments on the technological developments supporting the emergence of multi-role vessels in modern navies, including both manpower considerations and what modularity or flexibility means for the future cost-capability trade space for naval design. It refers to the future trends in naval surface ship designs and technologies, including expected trends in autonomy and unmanned systems. It challenges traditional views on deploying capability, considers the design and cost drivers for multi-role vessels and explores what this might mean in terms of how complexity impacts ship size and cost. A key consideration is how well the different roles can co-exist in one vessel, with examples given of different roles that would dovetail together well to address the concurrency requirement of different capabilities in a single ship. The paper reflects on the significance of naval manpower as the key through life cost driver for naval programmes, and the implications for the manning solutions and cost of future multi-role ship concepts.

1. INTRODUCTION

Since the end of the Second World War, many nations have had a changing landscape in the numbers and types of naval ships that have been procured and operated. Maintaining cost-effective, credible military capability continues to challenge naval forces, especially in recent decades as the pace of technological change in communications and weapons systems has developed such that today's advanced warships come with a very large price tag.

An issue for many countries is that the drumbeat of their naval shipbuilding has slowed in recent decades. Previously navies had several designs underway and one design followed closely behind another, however the design effort today is more disconnected between one generation of ship and the next. The trend of ships remaining in-service for longer naturally means mid-life updates become a focus, however modernisation of naval fleets using such updates is expensive and the architecture of the existing platform is a limitation. Designing adaptability into future vessels is therefore important, and increasingly the integration of modularity and flexibility into a ship design is therefore key [1].

One approach that may offer a compromise for acquisition agencies against the 'expensive, highly specialist platform' quandary is the potential to rationalise platform types within a fleet by increasing the suite of roles that each ship can provide. This is seen in the trend across several navies towards flexible, multi-role platforms that can enable a maritime force to deliver high-end capabilities across different scenarios, including non-combat operations such as Humanitarian Assistance and Disaster Relief (HADR). The inherent flexibility in multi-role ship designs allows them to stay abreast of the increasingly rapid development of technologies, which is attractive (if not essential) as navies attempt to keep up with such a trend.

2. EMERGENCE OF MULTI-ROLE VESSELS

2.1 HISTORICAL DEVELOPMENTS

The concept of a single ship being capable of doing more than one activity, or even of being employed on different missions, is of course not new. Even the frigates of Nelson's era would be capable of different roles: patrol or enforcement of blockades, 'scouting' activities in support of a main fleet, providing signalling (the communications of the day), landing small amphibious forces and anti-surface warfare against vessels of their own size.

The discussion in this paper is however related to the developments in modularity and flexibility that has led to the evolution of true multi-role ships over the past four decades. Examples of naval platforms illustrating this evolution are shown on the timeline in Figure 1.

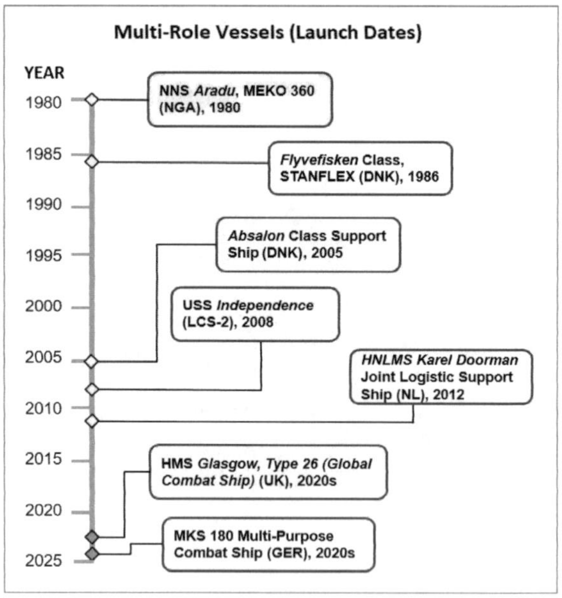

Figure 1: Timeline of Multi-Role Naval Vessels

2.2 MODULARITY OR FLEXIBILITY?

There is a need to define what we mean by the terms modularity and flexibility in the context of this paper. The most appropriate definitions are is given by [1] and are as follows:

Modularity involves creating fixed boundaries, defined interfaces, and defined ship services (such as power and cooling) to standard portions of a ship, which are termed modules. These modules can be plug-and play, such as the Danish STANFLEX concept seen on the Absalon Class ships (Figure 2), or fully modular installations of hotel services (such as galleys & medical facilities) that can be used across various classes of ship.

Flexibility involves the ability to change boundaries, whether they are physical or related to ship services. In the simplest case, flexibility can be provided by additional space margins in a ship, or the addition of extra capacity in ship's services within a particular compartment to allow for future upgrades.

2.3 EXAMPLES IN MODERN NAVIES

Current and future examples of multi-role ships include:

- The Danish Absalon Class Support Ship;
- The Dutch HNLMS Karel Doorman Joint Logistic Support Ship;
- The US Navy's Littoral Combat Ship (LCS);
- The Royal Navy's Type 26 Frigate (T26);
- The German Navy's Multi-Purpose Combat Ship (MKS 180).

A brief description of the Absalon and Karel Doorman classes, as examples of a smaller and larger multi-role platform, are given below:

- The Absalon Class (Figure 2) are a novel hybrid between a frigate and an assault ship, with flexible and modular capabilities allowing them to be to be equipped for naval warfare, land attack, strategic sealift missions or as command platforms. They can also be configured as medical ships or for HADR operations [2];

- The HNLMS Karel Doorman (Figure 3) is a joint logistic support vessel jointly operated by the Royal Netherlands Navy and the German Navy [3]. She provides amphibious assault capabilities with two davit launched Landing Craft Vehicle & Personnel, aviation facilities (hangar for up to 6 medium weight helicopters), medical and maintenance facilities and a 2,350 m^2 roll-on roll-off cargo deck.

Figure 2: Absalon Class Combat Support Ship (Naval Technology, [2])

Figure 3: HNLMS Karel Doorman (Government of the Netherlands, [4])

2.4 EXAMPLES OF RELEVANT NDP WORK

Naval Design Partnering (NDP) has been involved with a wide range of projects since 2011, some of which relate to the modular/flexible roles discussed here, including:

- Future Surface Combatant (FSC) mission bay (pre-Type 26 Concepts);
- NATO mission modularity studies;
- Medical Role 3 (MR3) Ship;
- Queen Elizabeth Class (QEC) Aircraft Carrier feasibility studies;
- Mine-countermeasures and Hydrographic

Capability (MHC) studies, with an Off-Board Systems approach to future capability;
- Navy Command MARCAP Multi-Role studies.

In 2017 NDP were tasked by the Maritime Capability section of Navy Command Headquarters (MARCAP, NCHQ) to investigate some multi-role feasibility studies into vessel types that could undertake various functions. The primary three roles NDP were asked to consider were:

- Host and transport an Amphibious Embarked Military Force (EMF);
- Incorporate a Forward Engineering Support capability;
- Host and sustain an Afloat Medical Capability.

Later in this paper we will use this example project to draw out discussion on multi-role design considerations.

3. FUTURE TECHNOLOGIES AND TRENDS

3.1 OPERATIONAL AND POLITICAL TRENDS

The evolving state of the 21st Century, certainly based on the first two decades, shows that the macro global trends will require states to be able to respond to both emerging political and conflict issues, as well as natural disasters (probably with an increasing frequency of occurrence and magnitude due to factors such as climate change, growing populations, etc.). The changing threats, at both low- and medium-level, mean that navies will need to be able to respond to both other nations and non-state actors, but additionally the more advanced navies will still need to be able to operate across a diverse range of warfare areas and with high-end capabilities [1].

3.2 TOPSIDE DESIGN

The development of increasingly sophisticated sensors and weapons will continue to drive topside design choices for high-end warships. In the medium term, the need for reduced Radar Cross Section design of the topsides is likely to continue, especially as more sophisticated approaches to avoiding detection and engagement will be needed against emerging technological threats. Future weapons, such as the US Navy's Railgun and the UK MOD's Dragonfire programme, will change the enablers and constraints on topside design, which will therefore need to evolve accordingly. Additionally, the anticipated reliance upon UXVs and the network enabled environment will make the Electromagnetic Compatibility (EMC) requirements on upper deck transmitter/receivers more complex.

The emergence of the integrated mast (shown in Figure 3), that allows the exploitation of modern materials and technology to improve sensor performance and coverage, has already had an impact on how naval ships are designed, integrated and operated. An integrated mast allows for a modular fit, where extensive shore-based testing can be undertaken prior to installation, simplifying commissioning procedures [5]. The next evolution may be the development of intelligent superstructure arrangements, with multiple modular sections providing real estate for sensors and communications systems.

Figure 4: NDP Previous Study examples (clockwise from top left): FSC Concepts / Mission Modularity / QEC / Medical Role 3 Concepts Ship

3.3 ONBOARD AUTONOMOUS SYSTEMS

A key driver for 'near-future' platform designs (i.e. those in the next decade or so), should capitalise on advances in the automation of manpower intensive tasks that are dirty, dull, deep and dangerous. Functions such as fire suppression, flood control, ship husbandry and the adoption of Condition Based Maintenance [6] are areas where automation will enable navies to reduce the manpower needed to deliver capabilities and therefore reduce Through-Life Costs (TLC) (see manning discussion below).

Figure 5: Topside and Spatial drivers

3.4. UNMANNED VEHICLES

The plethora of unmanned vehicles (UXVs) – or Off-Board Systems (OBS) – that are currently becoming 'the norm' as part of ship design and within new naval programmes means that integration of such assets into the mother ship is necessary. Factors such as space, weight, power, launch & recovery (LAR) systems, communications and control (which links to the previous point about topside design and EMC) all bring constraints to the ship designer.

We are now seeing the use of such unmanned assets to replace previously manned ones. Examples include the use of UAVs for over-the-horizon situational awareness by patrol and combatant ships [7], and the use of UXVs as part of future Mine Countermeasures (MCM) capabilities, as in the MHC Project in the UK. The latter approach reduces the human risk inherent in MCM operations by removing the personnel from the minefield, and sending Unmanned Underwater Vehicles (UUVs) to do the dull, deep, dangerous work instead. Multi-role vessels will of course need flexibility to host different UXVs, depending on the ship's operational role, and so the selection of suitable LAR systems is important.

The author recognises that UXVs need not necessarily be 'large and complex': future strategies may utilise swarms of low cost UXVs that can overwhelm traditional defences. It is also prudent to remember that a future platform will have to be capable of countering the same kind of threat. UXVs have the potential to be a game-changer which could overwhelm some existing capabilities and make them obsolete, just as the relatively low cost advent of maritime aviation in the Second World War did to the much more established and expensive Battleships and Battlecruisers [8].

3.5 NETWORK ENABLED ENVIRONMENT

The trend towards greater numbers of UXVs, remote systems and high-tech equipment on the battlefield is producing an increasing demand upon communication systems, and the power, bandwidth and coverage required to provide sufficient capacity. The increased levels of autonomy and sensors required for monitoring of on-board systems will similarly increase demand on the internal ship's network.

A future platform must have a communication and processing capability suitable for controlling, and collecting outputs from, all of the deployed assets (manned and unmanned) in a data-rich environment. It will need to be effective and affordable, and be reconfigurable and interoperable with allies, for example within NATO [9] (see Figure 6). Partly this can be achieved by using an open architecture approach with standardised interfaces, negating the need to re-engineer control interfaces and reducing the training burden for operators and maintainers [8]. This approach can also help to lower TLC and avoid early obsolescence of legacy systems, although the commercial and security issues have to be addressed effectively for this to be realised on naval platforms.

To ensure through-life flexibility, sufficient margins must be provided with regards to power, space and bandwidth, specifically to compartments likely to require upgrade during a vessel's lifecycle [10]. Examples would include the command facility or any compartments attributable to the control of UXVs.

3.6 MARINE ENGINEERING DEVELOPMENTS

The evolution of marine and auxiliary systems continues to play a part in how ships are designed. The adoption of an Integrated Full Electric Propulsion (IFEP) architecture, where the same generators can provide power for propulsion, ship's services and other capabilities, is one example of how design choices can impact on the flexibility achieved within a multi-role ship. The flexibility of layout that can be achieved (e.g. prime movers positioned independently of the shaft-lines, generators sited higher up to reduce uptake lengths, etc.) and the ability for power to be diverted to future energy weapons or sensors are just two points to note. Diverse power supply location also improves survivability, specifically reducing machinery space sizes.

IFEP architecture also enables future upgrade of a design, with future-proofing assisted by the power station concept and the possibility of more power for new systems. The developments in energy storage technologies will also support different architectures and systems, enabling new capabilities for pulsed loads, improved spinning reserve, reduced signatures and loiter drive.

The requirements for a loiter capability, where Amphibious or HADR operations may require a ship to be moving at slow speed in a littoral environment for a significant amount of time, can also be provided using hybrid propulsion options, thereby avoiding running the main prime movers at low loadings.

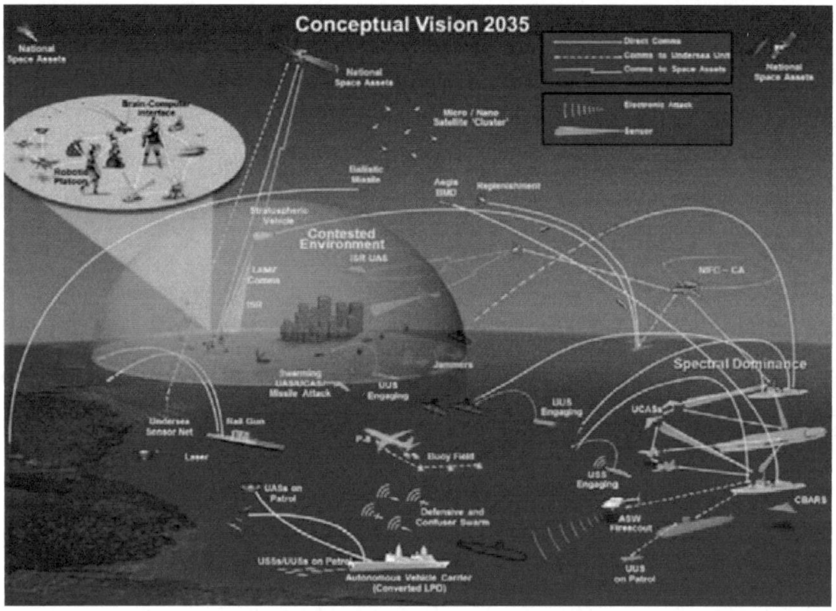

Figure 6: Conceptual Vision 2035
(SPAWAR (US Navy), [9])

4. NAVAL MANPOWER

4.1 SIGNIFICANCE AS A DRIVER

Manpower is a key consideration in any future ship design, with complement size being a primary factor that influences the overall demand on internal volume, hotel services and subsequently the Unit Production Cost (UPC) of the platform. It should also be noted that in the region of 80 % - 90 % of total Through-Life Cost is 'locked in' where these decisions are made early in the design lifecycle [11], and as much as 20 % of these TLC are directly attributable to operational manpower [12].

4.2 IMPLICATIONS FOR MULTI-ROLE SHIPS

The current indicators are that future platforms will become increasingly reliant upon automation, Artificial Intelligence (AI) and unmanned systems, noting that the extremely fast-paced development and adoption of such technologies is becoming increasingly apparent in both the civilian and military spheres. Currently however, the adoption of UXVs in the battlespace can paradoxically lead to an increase in the manpower demand placed upon the platform; and there have been naval examples of this [1]. Partly this is because of the large number of maintenance, operator and support staff required, but it should also be noted that the desirability of using human judgement to make key decisions is also a major factor. Future advances and increasing levels of 'trust' in AI, brought about by increasing acceptance in society, may help to reduce this burden [8].

The objective for future platforms (and especially when considering multi-role designs), should be manning levels that are 'appropriate' rather than minimal, achieved by utilising mature technologies and automation to reduce the burden upon the complement. Key lessons have been learned from recent programmes, such as the US Navy LCS [13], where un-validated assumptions to reduce crews had a negative effect on the material state of the platform. Rather than generating a cost saving, through reduced manpower, this actually led to increased costs from maintenance work. Ultimately it required an increase in crew size and therefore a reduction in the space available onboard for future flexibility.

4.3 MANNING STRATEGY

It should be remembered that for multi-role vessels the concurrency, or otherwise, of roles is a key aspect of the manning strategy. It is therefore important, that in fitting several mission capabilities into a platform, that the manning options are considered carefully at the outset. Choices about modular 'fit and remove' mission capabilities, with specialists that need to be embarked and disembarked effectively, versus the concurrency that can be achieved with permanently embarked multi-skilled operators, need to be considered. The manning strategy therefore directly drives the quantity and quality of accommodation to be provided within the platform.

4.4 NON-TRADITIONAL DEPLOYMENTS

Historically, a warship is designed for a single role, or at least a specific type of mission, and so the generation of a new design is based primarily on developments of that type of ship from previous experiences. Such evolutionary development used the military design standards relevant to the ship type, and assumed the traditional concept of a 'standard' crew: a set of highly trained naval personnel, formed into a complement that would deploy with the ship to go and carry out that role or mission.

Multi-role vessels, however, have the option to function in a very non-traditional way, and the historical view of 'the deployment' of a ship and her crew, going forth with a 'fixed' capability, will increasingly no longer be true.

Fleet planners in the future, using multi-role vessels, could have greater flexibility in how they schedule their ships to achieve their overall naval commitments and wider defence goals. Planning of future exercises to cover different types of emergent operational tasking should be much easier if the mix of platforms available come with multiple roles, and therefore can be 'swapped' between different deployments and missions.

The considerations of such 'innovation' needs to consider the holistic issues in changing how a navy operates. In the UK MOD, the approach to capturing such factors is through attention to the Defence Lines of Development (DLODs). These are:

- Training;
- Equipment;
- Personnel;
- Information;
- Concepts & Doctrine;
- Organisation;
- Infrastructure;
- Logistics.

5. ROLES AND MISSIONS

This section contains some examples of roles that have been considered by the NDP team, and common to several design studies for multi-role vessels, and which serve to highlight discussion points of some of the generic features of typical roles and capabilities.

5.1 AMPHIBIOUS

Amphibious operations have been supported with dedicated ship types and increasingly sophisticated surface connector types since the Second World War, enabling the projection of military power from the sea onto the land. Heavy military vehicles and equipment may still need to be taken ashore via surface assets, but in recent decades the need for medium/heavy helicopters has extended the range and shorted the timescale for an amphibious assault, making aviation facilities a key capability.

The Amphibious specialist ship type would include, for example, the Royal Navy's Albion Class (Landing Platform Dock) or the US Navy's Wasp Class (Landing Helicopter Dock).

In order to support an amphibious operation, the main drivers are the Flight Deck (for amphibious lift), capability for surface connectors (for example a Well Dock), accommodation for the EMF, space to carry vehicles (size of Vehicle Deck) and ramps for embarking them in a port. The 'flow' of personnel and vehicles to the aviation or surface assets for the amphibious assault is a key arrangement driver.

Figure 7: Drivers – Aviation, Surface Connectors and Containers (HADR/modular capabilities)

5.2 HUMANITARIAN ASSISTANCE AND DISASTER RELIEF

HADR operations require a ship with volume for carrying aid, such as food or emergency shelters, and the ability to transfer it effectively into crisis areas. The attributes of amphibious ships – specifically good aviation facilities and the ability to put logistics over the shore using facilities for amphibious surface assets – make them very suitable for the HADR role. An example of HADR operations include hurricane relief by the UK Bay Class (Landing Ship Auxiliary Dock). Effectively, such amphibious ships already provide a capability (albeit not tailored) to support such HADR missions as a secondary role.

As well as the spatial, aviation and surface connector drivers, other "transfer ashore" capabilities that HADR operations may need include fresh water and power, meaning generation of these may place system or spatial requirements on the multi-role designer.

5.3 FORWARD ENGINEERING SUPPORT

Forward Engineering Support in this context means deployed equipment and personnel that are able to conduct operational maintenance and repair for vessels whilst they are also deployed on operations. Similar to the changes already discussed in other naval areas, it is expected that the emerging trends on future enabling technologies will continue and result in developments in support and repair activities.

Previous repair ships generally had large workshops and storerooms of parts, along with specialist engineering personnel who could provide Forward Repair capabilities for other vessels. Today, the manpower required for Forward Engineering Support utilises fewer, highly specialised personnel. This contrast is shown by the example of between up to 700 repair and maintenance staff on HMS RESOURCE in the mid-20th Century, to the approximately 150 deployed on RFA DILIGENCE in the early 21st Century [14]. The workshops and accommodation areas were the previous spatial drivers.

As the technologies used by modern navies evolve, the maintenance footprint will continue to evolve with them. One area of development is the use of advanced and additive manufacturing technologies, which are already bringing changes in parts of the commercial maritime world [10]. The future of Forward Engineering Support could see a step change, away from large workshops and stores, to a deployable capability with advanced computational modelling power, libraries of spare part designs, and the basic raw materials for an additive manufacturing ability to 'replicate' broken parts.

Figure 8: Aviation (manned and unmanned) is a driver for many roles, including Amphibious and HADR

5.4 MEDICAL AFLOAT

Navies have to be prepared to deal with causalities in both peacetime and conflict. Although most naval ships have some capability to triage and treat casualties, for larger volumes of injured personnel a forward deployed Medical Facility is required. Where no host nation support is available to enable a land-based solution, an Afloat Medical Facility is required. An example of such a capability is the UK's Primary Casualty Receiving Ship, currently RFA ARGUS, which provides a deployable "Role 3" full hospital capability.

The main drivers for the Medical Afloat role are flow of casualties (i.e. main hospital facilities on the same level as the Flight Deck, to quickly transfer casualties arriving by air), accommodation for embarked medical personnel, and dedicated power, fresh water and HVAC requirements (which can be drivers for capacity of systems).

5.5 STRATEGIC SEALIFT FOR LAND FORCES

Large scale global deployments often require the sealift of land forces to be deployed to the theatre of operations. A number of navies have dedicated strategic sealift vessels for this purpose, but the function of a multi-role logistics/strategic lift is possible from an amphibious platform. This is because much of the infrastructure required is found in the Vehicle Deck, routes and ramps to associated surface connectors (even if the primary design intent was focussed on an amphibious assault, and so capacity is unlikely to match that of a dedicated platform).

The drivers for logistics and strategic lift are obviously the volume for Vehicle Deck or 'garage' areas, related to the weight and space constraints of the equipment or vehicles to be transported, with associated vehicle ramps and interfaces for offload.

6. DESIGN & COST DRIVERS

6.1 MULTI-ROLE DESIGN METHODOLOGY

The previous multi-role design example from NDP (see 2.4) included a conceptual scenario that was aligned with a future vision that several ships, ostensibly based on a single hull form, could be deployed around the world in different roles, but capable of relatively quickly being re-allocated to a different operation and with flexibility built in to enable a re-role without returning to home port.

The concurrency requirement for a multi-role platform, across several capabilities, will drive the spatial and systems requirements in the concept design. The ultimate extreme of this is 'total flexibility', whereby all capabilities (or roles) are fully realised, pushing the designer towards the expensive corner of the cost-capability trade-space.

The reality is that in naval design the cost-constrained design and acquisition environment is challenging, and budget constraints often lead to the design of any multi-role platform requiring compromise, trying to balance (often competing) requirements that the customer desires. It is important to recognise that a multi-role platform cannot be expected to achieve the same level of performance in an individual role as a platform that was designed and optimised solely for that purpose. The key is to try to strike the correct balance of cost, capability and flexibility so as to provide the operator with the most useful platform possible at a sensible cost.

Traditional ship cost models see a larger ship (more volume) as more expensive, however a larger vessel may bring TLC savings. It is recognised [1] that one of the future flexibility features will be 'volume margins' in naval ship design: a bigger ship can be cheaper through-life, as it enables upgrades and future capabilities to be added.

6.2. CONCURRENCY AND A BIAS APPROACH

In the previous NDP design example, work considering multi-role vessels with Amphibious and Medical Afloat roles identified early on that these requirements would be key spatial design drivers. This was because of the dedicated equipment and utilities (power, HVAC, fresh water) requirements of the Medical Afloat role versus the alternative large EMF accommodation requirement for the Amphibious role (which had much lesser demands in terms of systems). An option to have 'total flexibility' to be able to rapidly reconfigure between the associated two sets of functions was quickly identified as going to drive a very large ship indeed.

NDP therefore adopted a 'Tailored Flexibility' approach for that project, where vessels in the class used a common hull but have a bias either for the Medical Afloat or the Amphibious role. This approach removed the need for all the ships in the flotilla to have the capacity to accommodate both the full Amphibious and Medical Afloat requirements, while still benefiting from a large degree of commonality between the vessels.

Clearly the flow of personnel (amphibious troops, evacuees or casualties) and the onboard accommodation requirements for them, together with aviation or surface connection assets, link the Amphibious and Medical Afloat roles quite closely (and HADR if a required role), suggesting the drivers of such capabilities can dovetail together well in a multi-role design.

The NDP study therefore proposed concepts with a flexible spatial arrangement, necessary to enable the internal compartments to be rearranged as needed to support the new functions required [1]. Similarly, along with this was the operational need to accommodate the 'Embarked Specialist' personnel for the roles.

6.3 SIZE DRIVERS

The NDP study developed an initial sizing tool, based on deck area demands generated from the MARCAP requirements, in order to rapidly assess the vessel length drivers for the multi-role vessels. The tool considered 128 iterations of the key input variables to help identify the key length drivers and to inform the selection of a suitable hull form. The following assumptions were made in the configuration of the initial sizing designs:

- The vessel beam was driven by the Amphibious requirements for Well Dock and Vehicle Deck width, plus a margin for access either side;
- The Flight Deck would be positioned aft;
- The Hangar would be positioned immediately forward of the Flight Deck;
- The Superstructure (which includes the Medical Facility) would be positioned immediately forward of the Hangar.

A typical vessel solution generated by the initial sizing tool is shown in Figure 9.

The output of the initial sizing indicated appropriate vessel length within a set bracket, and with a matching minimum beam requirement. These indications / constraints provided the starting point for the principal dimensions of the first design iteration and informed selection of a suitable hull form for use in the concept model. More detailed design variants were then developed by building upon the data from initial sizing and using NDP's concept design experience.

The key design drivers for the NDP multi-role concepts included:

- Retaining commonality of design (structural and systems) across different bias solutions;
- Well Dock / Surface connectors;
- Vehicle Deck;
- Medical Spaces;
- EMF Accommodation;
- Flight Deck and organic aviation facilities.

The desire to have maximum structural commonality across the hulls in the intended flotilla meant that an early decision during initial sizing was to fix an architecture including a Well Dock aft. For one of the more onerous amphibious operational scenarios considered, this meant the organic surface connectors could be carried 'equally' across all the example ships in a Task Group. Initial sizing considerations for the Well Dock operational concept stemmed from a separate NDP LAR sub-study within the project: it is important to realise individual aspects of a multi-role design may need specific studies to evaluate if factors are a driver or not.

6.4 MULTI-ROLE COST DRIVERS

The factors driving the cost of a ship are naturally linked to those that drive the design, as discussed above. This is both in terms of the final output (i.e. the design that is built and hence the UPC), but also the Non-Recurring Expenditure (NRE) during the design and approvals cycles. The latter, especially for naval ships where the scrutiny by government and political entities is required due to the commitment of public funds, can of course be quite significant.

Figure 10 illustrates the "cost onion" for the Personnel cost drivers, linked to the previous manning philosophy discussion. This shows that, although the Pay element is obviously a direct function of employing personnel on a ship, this is only a small aspect of the overall TLC of Personnel. It further illustrates the complexity of the second and third order effects that span across the DLODs.

NDP, like other government concept organisations, have bespoke and detailed cost models and metrics to evaluate the impact and significance of how cost is driven by design changes. These models cannot be detailed here, but clearly the overall cost is driven by factors such as:

- Complexity of command systems and sensors;
- Manning philosophy (including the "Personnel Cost Onion" drivers);
- Size of accommodation systems (both the quality of accommodation and the ship services supporting it).

7. CONCLUSIONS

This paper has discussed the attraction of multi-role vessels for navies who need cost-effective options for procuring platforms. The future global trends would appear to indicate that navies will continue to have a need to deliver high-end capabilities across different scenarios and be able to react quickly to emerging political scenarios and natural disasters.

The Multi-Role case study presented from previous NDP work demonstrates how a viable pre-concept solution for a multi-role class of ships can be developed to support at least three primary capabilities, with varying amounts of concurrency in the functions delivered at any one time. The example demonstrates that understanding the concurrency levels of the different roles and the key requirement areas, such as use of automation, manning and embarked assets (surface connectors, UXVs, etc.), that impact key ship sizing drivers, is crucial to achieving a feasible solution.

How well the different roles co-exist in the requirements of the multi-role concept are an indication of the level of complexity that will be involved in developing a solution; this complexity then directly influences the ship size and cost that will be arrived at. Amphibious, Medical Afloat and HADR capabilities are a good example of a set of different functions that can dovetail together well in a multi-role platform. The accommodation levels required for embarked 'mission specialists' will be a significant ship size driver, so a realistic concept of how manpower will be deployed is necessary to inform this and prevent the vessel being over-sized. The designer must have this overall manning strategy for a multi-role ship as an input to the design process, if the solution is to work effectively with the core crew / mission specialist mix to be embarked.

Current indicators are that future platforms will become increasingly reliant upon automation, AI and unmanned systems, noting that the extremely fast-paced development and reduction in cost of such technologies is becoming increasingly apparent in both the civilian and military spheres.

The evolution of UXVs is likely to lead to a step change in how military operations take place in the future and this will include different naval doctrine and associated ship design developments. The author suggests that the evolution of manning philosophies alongside these technologies will be a key enabler.

Future multi-role ships will undoubtedly be required to interface with an ever larger range of UXVs supporting different naval operations (so, for example, the flexibility of LAR systems will be important). The trends in UXV technologies (and the ship designer's adoption of modular or flexible solutions to incorporate them) will be key to the evolution of multi-role shipping in the decades to come.

It is recognised [1] that one of the future flexibility features will be 'volume margins' in platform design: enabling a bigger ship to be cheaper through-life, as upgrades and future capabilities will be able to be added more efficiently. This suggest that a suitable challenge is needed to the traditional "size = cost" viewpoint and a shift into accepting that a larger ship will enable future flexibility and easier upgrades through-life.

8. ACKNOWLEDGEMENTS

The author would like to thank the management team at NDP for enabling him to include in this paper the insights into future designs from NDP concept work, and also to our Customers at the Maritime Capability (MARCAP) section, Navy Command HQ, for their ongoing sponsorship.

I also acknowledge the work of my colleagues in NDP, who have undertaken multi-role vessel concept studies; especially Dave Evans for his skills with rendering such concepts into illustrations and Phil Henderson who led the initial sizing work referred to in this paper.

Disclaimer: whilst the support of NDP colleagues is recognised, please note that the opinions expressed in this paper are solely those of the author and do not necessarily represent the views and opinions of the Ministry of Defence or Babcock.

9. REFERENCES

1. Schank, J. et al., 'Designing Adaptable Ships - Modularity and Flexibility in Future Ship Designs', *RAND*, 2016.

2. Naval Technology, 'Absalon Class Combat / Flexible Support Ship', [Online] Available at: *https://www.naval-technology.com/projects/absalon/* [Accessed 03 06 2019].

3. Naval Technology, 'Karel Doorman Joint Logistic Support Ship', [Online] Available at: *https://www.naval-technology.com/projects/karel-doorman-joint-logistic-support-ship/* [Accessed 03 06 2019].

4. Government of the Netherlands, 'HNLMS Karel Doorman and army engineers set course for the Caribbean', 2017, [Online] Available at: *https://www.government.nl/latest/news/2017/09/14/hnlms-karel-doorman-and-army-engineers-to-set-course-for-the-caribbean* [Accessed 03 06 2019].

5. THALES Group, 'Integrated Mast', [Online] Available at: *https://www.thalesgroup.com/en/innovation/integrated-mast#* [Accessed 15 06 2018].

6. UK MOD, 'Joint Concepts Note 1/17 - Future Force Concept', *Development, Concepts and Doctrine Centre (DCDC)*, 2017.

7. CLARKE, M., 'Integration of Rotary Wing Capability into Surface Ships', *Future Surface Fleet Conference*, 2018.

8. UK MOD, 'Joint Concepts Note 1/18 - Human-Machine Teaming', *Development, Concepts and Doctrine Centre (DCDC)*, 2018.

9. SPAWAR (US Navy), 'Space and Naval Warfare Systems Command - STP Critical to Navy Space', 2017. [Online] Available at: *http://www.public.navy.mil/spawar/News/Documents/Presentations/05.17.2017_STP.pdf* [Accessed 10 06 2018].

10. DNV GL, 'New centre to boost 3D printing in oil and gas industry', 2018. [Online] Available at: *https://www.dnvgl.com/oilgas/perspectives/new-centre-to-boost-3D-printing-in-oil-and-gas-industry.html* [Accessed 15 06 2018].

11. US GAO, 'Navy Actions Needed to Optimize Ship Crew Size and Reduce Total Ownership Costs', *US Government Accountability Office*, 2003.

12. Rabbets, T., Webb, R. & Astle, J., 'Whole Life Cost Prediction and Reduction for Maritime Platforms', *MOD Report by Qinetiq*, 2003.

13. US GAO, 'Navy Force Structure: Actions Needed to Ensure Proper Size and Composition of Ship Crews', *US Government Accountability Office*, 2017.

14. Kimber, A., 'Future Ship Concepts for Repair and Maintenance at Sea', *World Maritime Technology Conference*, 2006.

10. AUTHOR'S BIOGRAPHY

Dan Smith holds the current position of Principal Engineer (Naval Architecture), within MOD's Naval Design Partnering Team, where he is seconded from Babcock. He is responsible for ship concept design and costing studies for the RN and RFA. His previous experience includes Naval Architecture roles within Rolls-Royce Naval Marine and the UK MOD.

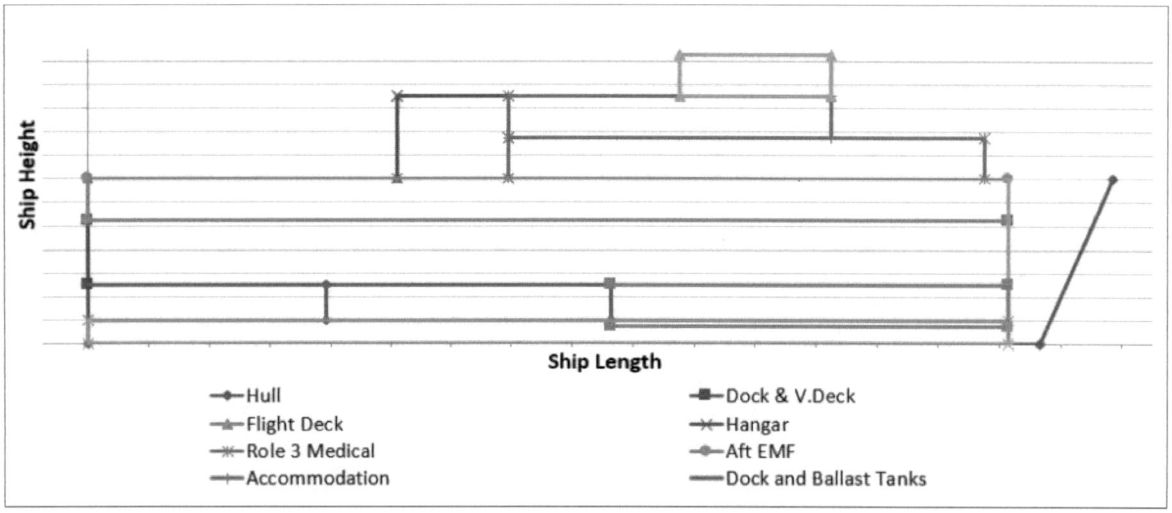

Figure 9: Initial sizing tool – profile view of a design iteration

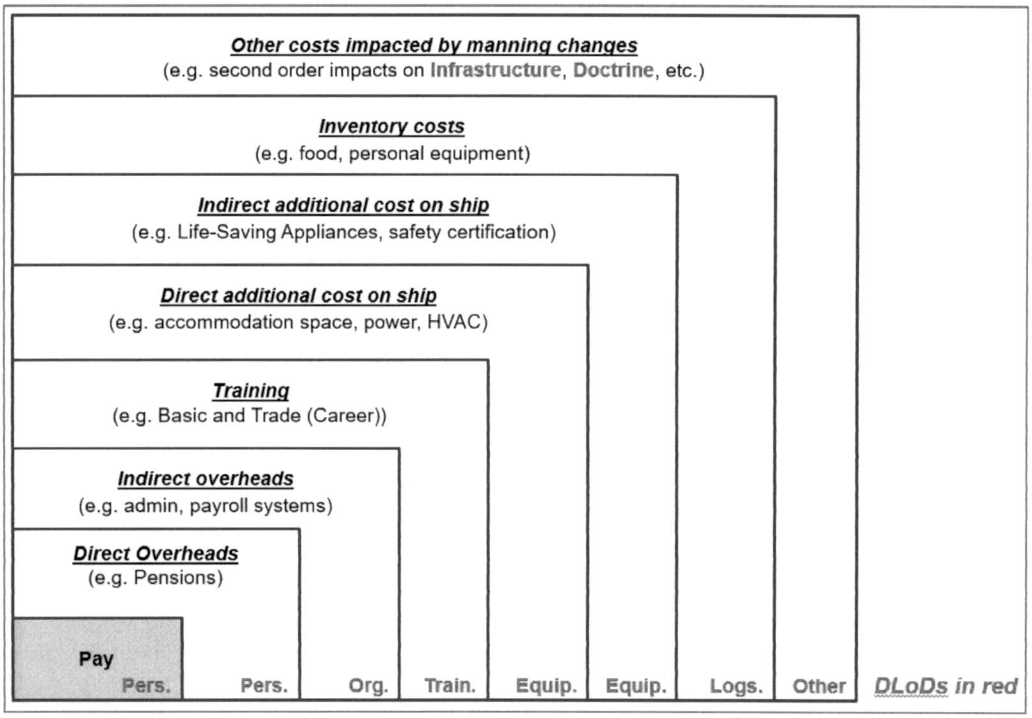

Figure 10: NDP's "Personnel Cost Onion"

THE MULTI ROLE COMBATANT – JACK OF ALL TRADES, MASTER OF NONE?

D Andrews, FREng, PhD, FRINA, RCNC, UNIVERSITY COLLEGE LONDON, UK

SUMMARY

For naval vessels Modularity is essentially a design style choice adopted to achieve both aspects of multi-role capability, namely, Flexibility (in deployment) and Adaptability (through life). This paper largely focuses on the second of these, which has been seen to be an attractive objective in Western naval vessel acquisition since the end of the Cold War. A paper to one of this series of warship conferences produced by the author a decade after the collapse of the Warsaw Pact was entitled "Adaptability – the Key to Modern Warship Design". It particularly drew on the difficulties the author had experienced in establishing the case for the Future Surface Combatant (the precursor of the Type 26 Frigate), where the primary capability of this first post-Cold War combatant as a globally deployable class of warship was to demonstrate that Adaptability would be good value for money in an uncertain geopolitical environment.

The current paper commences by revisiting the case for Adaptability nearly two decades on from that first consideration and whether the same drivers for Adaptability and advances in Modularity have made the case for multi-role combatants more convincing or whether they are still prone, if they are not truly "First Rate" designs, to be "Jacks of all trades, yet Masters of none?" Next the paper reviews past warship designs that have been inherently adaptable, leading on to consideration of "margins", the traditional way naval ship designers have been able to incorporate a measure of adaptability. However, non-ship design aware procurement authorities, have continued to squeeze seemingly generous margins as a short sighted means of achieving "savings" in the procurement element of defence budgets. There are of course other features, necessary to achieve Adaptability. Thus the paper explores how an adequate consideration of the ship architecture might provide an adaptable design and, specifically, how adopting Modularity to achieve this, can be explored early in ship design. Finally, leading on from the exploration of Adaptability, the paper concludes by a considering the Trimaran configuration. This configuration remains a highly attractive overall Style and ship configuration choice for future warships, precisely because it readily provides so many of the features that are seen to be appropriate in an Adaptable Multi-role Warship, which can be configured to avoid the Second-Rate "master of none" trap.

1. INTRODUCTION – ADAPTABILITY THE KEY CAPABILITY FOR MODERN WARSHIP DESIGN

The choice of a modular "Style" (Andrews 2017) for a naval combatant is both a highly significant design choice and also a direct physical solution to achieving a multi-role capability in a (potentially) satisfactory manner. That is to say in the most "cost effective" manner in a defence acquisition environment ever more constrained by budgetary pressures, particularly on warship procurement. The pertinent question, with the loaded term "cost effective" (see Hockberger (1993) for an analysis of the US Navy's adoption of Cost and Operational Effectiveness Assessment practice), is whether this is both effective in fleet and ship performance terms and, hence, value for money (VFM). The latter should be in terms of the naval budget, rather than just applying the usual key measure of individual ship procurement, namely (for UK vessels), Unit Procurement Cost (UPC), which itself is seen as a dubious basis for VFM (Andrews 2018a).

It could be said that Modularity is a Design Style solution to the issue of Adaptability, which has been an objective to the fore of naval ship acquisition since the end of the Cold War. It was thus addressed by the author in a paper presented to the RINA 2001 Warship Conference (Andrews, 2001), when it was seen to be the prime capability sought in new UK naval combatants in the uncertain world emerging after the collapse of the Soviet Union. At that time all existing naval vessels, including those in the shipbuilding programme under way, such as the Type 23 Frigates and the ASTUTE Class SSNs, had been conceived at the height of the Cold War when the prime direct naval threat was the massive submarine force built up by the Soviet Union. So multi-role vessels were not as appropriate and ASW was seen as the over-riding warfare domain, with AAD as a necessary adjunct to protect the main UK maritime forces in that primary ASW role. Thus for example both the Type 23 Frigates, as initially developed from specific operational analysis of intended towed array sonar operations in the GRIUK Gap, and the INVINCIBLE Class carriers, ordered as ASW helicopter carriers, were both only belatedly modified to have more general purpose or multi-role capabilities. So the Type 23 had a medium calibre 4.5inch gun added very late in the design (post the Falklands Campaign (Thomas & Easton, 1992)) and the carriers incorporated the (multi-role) FRS (Fighter, Reconnaissance and Strike) Sea Harrier (Honnor & Andrews 1982). The latter incorporation commenced in May 1975 two years into the build of the First of Class when it had become obvious that the RAF could not provide the naval air cover it committed to give. That commitment having been the basis of the 1996 decision to scrap the programme to procure the much larger fixed wing capable CVA01 Class carriers.

The 2001 paper on Adaptability identified two prime reasons why, post the Cold War, future naval vessels should have Adaptability as their primary capability driver:

a. The geopolitical environment in which future naval operations would be conducted was likely to be highly unpredictable, in marked contrast to the operational context in which the vast majority, if not the totality, of existing warships had been designed. The US Navy then led on the way forward by refocusing the operational stance of Western navies post-Cold War with the doctrine of "From the Sea" (Kelso, 1992). In the case of the R.N. the political redirection was spelt out in the Strategic Defence Review of 1998 (UK MoD, 1998). The SDR therefore emphasises the need for the UK Defence Forces to operate world-wide and across a broad range of Defence Missions. These range from, at the 'hot' end, full power block confrontation (as in the Cold War), through UN and NATO Peace Keeping to Defence Diplomacy and Constabulary tasks (such as countering the illegal drugs trade and piracy). This resulted in requirements for new warships being less amenable to precise Operational Analysis (OA) scenarios, together with an emphasis on "flexibility" in operations and the ability for the basic design to readily accommodate new roles and provide new capabilities in service. This was seen as being achieved through built-in flexibility and incorporating adaptable features to more readily accommodate introducing previously unforeseen/immature new roles and technologies. Needless to say this is a far more demanding design requirement than the previous approach, which was driven by highly tuned OA based requirements. In part this new stance led to the 1998 SDR strongly emphasising the need for all new major defence acquisitions to be designed for Technology Insertion (UK MoD, 1998).

b. The second driver, which did not really change with the end of the Cold War (it just got worse), was the budgetary squeeze on naval acquisition. In major defence procurement, almost from time immemorial, there has usually been a reluctance in a democracy to pay any more than the absolute minimum for a new military product. Until very recently this meant minimising the initial procurement cost ('maximising the Bang for the Buck') and very often at the price of increasing the running costs, including the ability to upgrade the new product through out its operational life. This can be seen at an extreme for naval vessels, which as the largest of complex individual military units, have highly politically visible procurement costs. With the end of the Cold War, the victorious democracies looked to the Peace Dividend to reallocate national resources away from defence, and so concurrently with the perceived need for less highly tuned warships was the pressure to reduce the cost of warships, however that is quantified. This has occurred despite the recognition that the through life costs, rather than just initial procurement costs, ought to govern defence equipment acquisition decisions. So alongside the reassessed national defence imperatives, warship designers had to refocus on the challenging issue of "Design for Adaptability".

This design intent is considered to be a significant challenge, at least in the case of warship design, because warships have historically been inherently flexible in their broad design intent and, pre-eminently in the Royal Navy, in their deployment as instruments of national policy. For this reason, the next section reviews past warship designs that have been inherently adaptable. This leads on to consideration of "margins", which can be said to be the traditional way naval ship designers have been able to incorporate a measure of adaptability. However, procurement authorities, being largely unaware of the nature of ship design, have continued to squeeze what they consider to be generous margins, as a short sighted means of achieving apparent "savings" in the procurement element of defence budgets. Yet, adequate margins remain necessary if not sufficient to provide adaptability in a warship. In discussing the other features, which are seen to be required to achieve Adaptability, particular attention is focused on the architecture of the adaptable design and how well Modularity, as a means of achieving this, can be explored early in ship design. Finally, leading on from the exploration of Adaptability, the paper concludes with a reassessment of the Trimaran configuration for naval combatants. This configuration is seen to be a highly attractive overall Style option and ship configuration choice for future warships. That is precisely because it readily provides so many of the features that are seen to be appropriate in an Adaptable Multi-role surface combatant and thus able to avoid the trap of the second-rate “master of none” solution.

2. DESIGNS THAT HAVE BEEN INHERENTLY ADAPTABLE

This section looks at some specific warship designs with contrasting approaches to the way in which the need for adaptability has been achieved. The designs are largely of warships which have been in service in recent decades, rather than paper designs, and where often contrasting views have been expressed as to the degree to which they have meet their design intent and often, even, the validity of that intent. They are largely British designs, which have been described in RINA papers, but in addition for those designs reference has been made to Brown's RCNC history (Brown, 1983), which captured individual designers' comments often relevant to design for adaptability. All the designs are of the Destroyer/Frigate category, with one exception (The INVINCBLE Class Carriers), and there is also one non-UK designed example (The SPURANCE/ TICONDEROGA Classes) as it is a well-documented example of margin erosion over the life span of the basic design. The discussion presented focuses on the issue of adaptability and clearly cannot do justice to the individual designs in the round, nor fully capture the complexities of the many issues, other than adaptability, which may have been seen to have been more significant to their designers in many or indeed most instances.

2.1 FIRST AND SECOND RATE DESIGNS

In an early study into the nature of ship cost, Brown and Andrews (1980) drew attention to a series of R.N. ship designs to point out that certain ship classes, which had been specifically designated 'first rate' or 'second rate' designs, invariably showed that the latter were poor VFM. This applied to the First World War battleships of the Queen Elisabeth Class and the Revenge Class, where the latter 'cheaper' class were far less effective and clearly less VFM, as shown by the latter's inability to be upgraded over a thirty year life. In WWII the early convoy escorts, the Flower Class corvettes, were also poor VFM in that role compared with the later Castle and Loch class frigates. Immediate Post War, first and second class ASW frigates were produced (Types 12 and 14 respectively) and while the former led to the very successful LEANDER Class, the Type 14s were soon disposed of as lacking the flexibility in the Cold War era, after the Korean War. This pattern occurred, despite all these first rate designs being initially considered 'too expensive' and the second rates seen to be the means of increasing hull numbers for broadly the same primary role – that of ASW in the last example. All these comparative designs well exhibit the overall design style choice being made from which all the capabilities flow (Andrews 2017).

2.2 COUNTIES COMPARED TO TYPE 42 DESTROYERS

The contrast between the RN's first two classes of missile armed Air Defence Destroyers, namely the County and Type 42 Classes (Purvis, 1974), could not be more marked with regard to Adaptability. Although the Counties originated as large gun destroyers they were quickly redesigned to be the platform for the new large Sea Slug AAW missile system. They became a large comfortable design, which was configured with particular attention to generously provided ship's services, an essential feature if a design is to be adaptable through life. This ability was tested late in the design rather than in service, when the requirement to add a hanger and flight deck space for the not inconsiderable Sea King helicopter was imposed by the Admiralty Board. That this could be done, and incidentally several of the class late in their life could be significantly modified to carry several Sea Kings by the Chilean Navy, demonstrates that a sensibly designed and generously margin provisioned ship can be highly adaptable. The Type 42 Destroyer design was in stark contrast, in that as a clear victim of being 'designed down to a price' (Purvis, 1974) there were no margins for future technology insertion or role flexibility. The very apparent short sightedness of this approach was revealed when the last batch of the class had a mid-body addition on build, to give at least the hydrodynamic capabilities their designer originally wanted (Brown, 1983). It is also worth pointing out that this restoration of a reasonable hull form did not change the rest of the design, which remained too congested to readily facilitate technology insertion and provide additional ship services.

2.3 TYPE 22 COMPARED TO TYPE 23 FRIGATES

A similar contrast can be drawn between the Type 42's ASW consort, the Type 22, and the latter's eventual successor, the Type 23 Frigate classes (Purvis, 1974, Thomas & Easton, 1988). Despite some initial constraints, the Type 22 was conceived as a proper replacement for the LEANDER Class as the main ASW Frigate in the fleet and was designed with healthy margins and a spacious layout. This initially resulted in senior naval criticism that it was an under-armed ship for its size, in a very similar manner to the criticism of the SPURANCE Class in the U. S. Navy when they were first introduced into service (see Section 2.6). However, the contrast in the relative ease with which the later batches of the Type 22s were stretched and considerably enhanced with vital new sensors, serves only to highlight again the short sightedness of some senior decision makers, with a lack of understanding of ship design. In the case of the Type 23 Frigates, the responsibility for yet another design "without any margins" lies in a combination of the unusual origin of the design, not as the Type 22 Replacement but as "a towed array Tug", plus the extremely fraught political environment in which the design evolved. The operational requirement was also justified, more so than for earlier designs, by the extremely specific OA modelling of TA operations referred to in Section 1. That OA on which the ship design drew was then made redundant by the fact that the Cold War ended before the First of Class entered service. The final flaw in the overall procurement concept for the Type 23 was that the design should be marginless, which was directly attributable to the top-level decision to make this class have a service life of only 18 years. The fact that the remaining ships of the class are, currently, looking to some 30 years typical service life only reinforces the need for intelligent ship design knowledge to be given reasonable weighting in top level procurement decision making. In the case of the Type 23 short life decision all those involved at the concept design level knew that this decision would not be adhered to but were unable to convince the policy makers to restore adequate margins. The through life cost implications can be squarely laid at the altar of political interference in ship design matters.

2.4 R.C.N. ST LAURENT CLASS

A contrasting design to the "marginless/non adaptable ship" is provided by Baker's St Laurent design produced when he was in Canada in the 1950s (Baker, 1956). He not only provided adequate margins in the design, which enabled the ships to subsequently accommodate a very large helicopter on a relatively small ship, he also adopted a "Stylised" layout. This clearly delineated the various compartments in a given category (i.e. Payload, Living, Working, Services, Liquids) from the other categories, thus ensuring, as in the case of the Counties above, that the considerable inboard impacts of subsequent technology insertion were achieved with minimum disruption to the rest of the existing ship fabric. The design, like the Type 22, was readily modified in later batches to accommodate enhanced weapon fits. It is also relevant that, with a more generous beam than the contemporary British designs (Brown, 1983), the inevitable degradation in statical stability righting lever, due to top weight growth, was more readily accommodated, thus justifying Baker's design insight and strong design control.

2.5 THE CAH INVINCIBLE CLASS CARRIERS

An example of an adaptable ship at the top end of the warship spectrum is that of the INVINCIBLE Class Carrier. The concept of this design developed out of that of an ASW Helicopter Cruiser, which because of its large helicopter complement and the clear intent to make its through deck and hangar able to operate VSTOL aircraft, was quickly modified into a small but genuine aircraft carrier (Honnor & Andrews, 1982). Large warships designed to carry and operate several aircraft or amphibious forces and vehicles have, by virtue of the volume and deck area required, inherently adaptable configurations. Thus the INVINCIBLE Class have been modified over the years to accommodate considerable enhancements in their aircraft, combat suite and complement. This has been due principally to three features consciously incorporated in the design. Firstly, generous margins, not just in stability but also in structural and ship services standards. Secondly the General Arrangements exhibit a sophistication in balancing the many demands of general access, storing, removal routes, escape routes, weapon handling, command and control, aviation support and damage control with the major constraint of a long hangar and flight deck – see Figure 1 (Honnor & Andrews, 1982). Finally from the start of the ship design, the major aviation features (hangar and lift dimensions, hangar and flight deck structural design) were designed with technology insertion in mind. This was done by avoiding the expedient of just designing for the current generation of aircraft, rotary and VSTOL, but rather for the expected next generation of both types, including allowances in the ship's structural scantlings and generating capacity. These seemingly generous margins have enabled the adaptation and useful life of these vessels to be extended and this philosophy has been also been adopted in the RN's Future Carrier, despite its main design intent being to maximise sortie generation rate as the prime requirement (Knight, 2009).

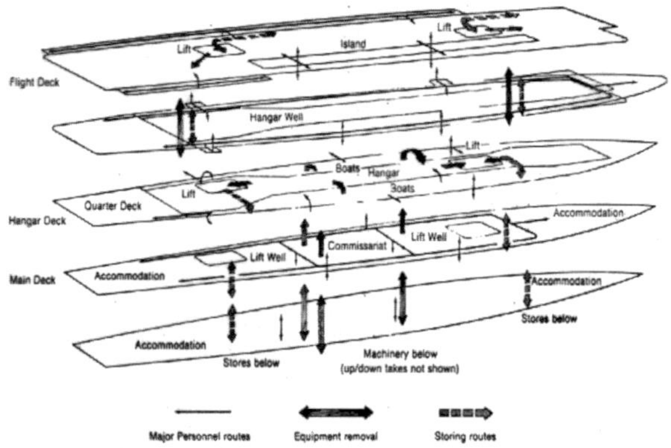

Figure 1: The INVINCIBLE Class Carrier Layout Logic (Honnor & Andrews, 1982)

2.6 U.S.N. SPRUANCE/KIDD/TICONDEROGA CLASSES

The final example of an adaptable design is that of the U.S. Navy's DD963 or SPRUANCE Class (Leopold & Reuter, 1970). This whole class was procured under the Total Package Procurement policy of Secretary of Defense McNamara, with a strong through life support incentive on the contractor, Littons. Leopold therefore built into the design not just very generous margins but also left significant portions of critical compartments, such as the CIC (Combat Information Centre, namely the Operations Room in UK parlance), empty and labelled as "Reserved for future equipment". These ships were initially criticised in the fleet as being under armed at 7700 tons deep displacement. However, the design proved to be highly adaptable such that a more heavily armed version (initially for Iran then incorporated in the U.S. Navy as the KIDD Class) completed at 9200 tons, and the ship design was then used for the Aegis Cruisers, CG 47 or TICONDEROGA Class. However, as Table 1 shows, this final extrapolation was not in the end a success, as the first of this latter class was completed without the equivalent of the UK Board Margin (see Section 4) thus having "no margins for future characteristic changes" (Watkins, 1982) and too close to its structural and damage stability limits at the start of the design's service life. This sounds very like the fact that the CG 47 team, in focusing on the high-risk development of the Aegis AAD component of the project, had forgotten to build in even minimal adaptability to the ship. Thus the lead combat system engineers did not address the whole system of systems that constitutes a warship (Andrews, 2011).

Table 1: Chronology of Design Growth in a US Navy Combatant Design (Watkins, 1982)

SPRUANCE			
	on completion(1)	Displacement	7700 tons
		Complement	240
		Accommodation	296 (20% growth allowance)
	growth	Displacement	8000 tons
	'in a few years'(2)	K.G rise	0.3 ft
		Manning	36% increase
KIDD			
	on completion(3)	Displacement	9200 tons
TICONDEROGA			
	contract target(3)	Displacement	9200 tons
	on completion(3)	Displacement	9500 tons
	with minor payload additions(3)	Displacement	9600 tons
	'naval architectural(3) limit'	Displacement	10,200 tons

Notes:

1 Collins (1975)
2 Keane (1982)
3 Watkins (1982)

3. MODULARITY CASE STUDIES

As early as 1971 the then Advanced Ship Concept Team in the US Navy's NavShips (now NAVSEA) in a paper entitled "A Philosophy of Naval Ship Design and Construction" examined the "broad construction philosophy of modularity" (Spero et al, 1971). Figure 2 taken from that paper shows the concept for "single mission escort ships" with exchangeable "functional payloads" for AAW, ASW and shore bombardment. This was a very putative concept and a much more substantive US version of Modularity was shown by Broome & Friedman (1992) looking generally at Future Combatants. Figure 3 from their paper is entitled a Variable Payload Ship, which they describe as designed "for ease of change during construction and service life". They remark that such vessels are likely to be larger than present designs, however "sheer size does not entail higher cost". This does not seem to have been borne out as the current set of combatants have got larger without any noticeable reduction in outfitting costs, which still seem to be priced on historic weight based initial cost (Andrews, 2018a).

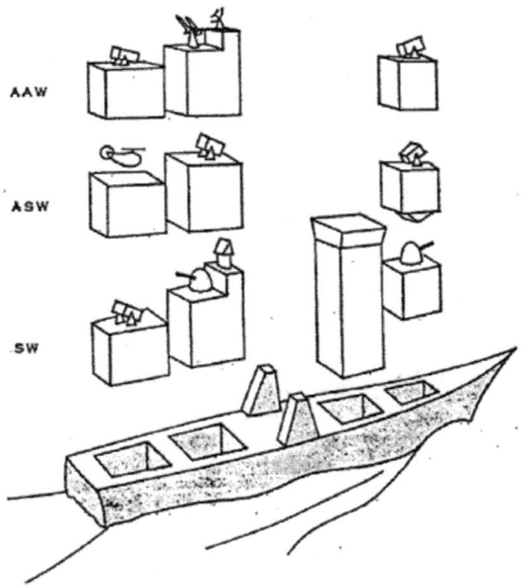

Figure 2: US Navy Modular Escort Ship Concept (Spero et al, 1971)

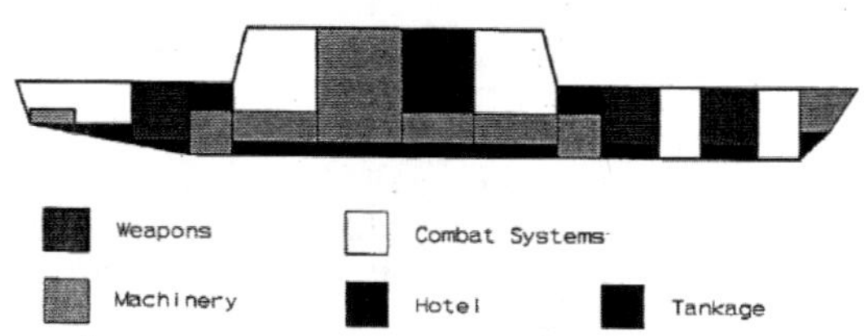

Figure 3: US Navy Variable Payload Ship concept (Broome & Friedman, 1992)

ARAPAHO (Betts & Harris, 1984) was a NATO project, which Modularised the provision in standard 40ft. T.E.U.s of all the requirements to operate and sustain 4 SeaKing ASW helicopters fitted to a containership to provide emergency convoy defence. The full capability provided of flight deck, hangar, accommodation, weapons, C3, servicing, on-board logistics and additional power, required 73 specially outfitted 40ft. T.E.U.s and 92 modules (for the hangar and flight deck structure). After the first ship set was manufactured and trialled at sea the follow on provision of sets for an adequate number of convoys to be protected was not proceeded with. However, the exercise clearly demonstrated that modularising (even with the physical convenience of attachment to container ship spigots) is extremely space inefficient compared to "traditional integration" of combat systems into warships. Proponents of Modularity are likely to respond that this can be balanced against the expense of outfitting naval combatants both on build and in any upgrading through life – however through life cost savings are less easily argued through defence acquisition bureaucracies and short term revenue focused politicians.

Two systems that are said to be modular that have gone to sea in significant numbers of hulls are the Blohm & Voss MEKO system and the Danish Navy's smaller (coastal craft) STANFLEX approach. Both systems were pertinently discussed by Figg (1994), in addressing "Affordable Warships". The MEKO system concentrates on "modularising" major weapon equipment (such as whole gun emplacements), provoking Figg's comment "little thought is given to (ship & propulsion) systems being integrated". STANFLEX has been subsequently adopted in larger ocean going vessels (e.g. HMDS ABSOLOM) where a large mission bay can more readily accommodate different modules from the STANFLEX range.

The UK Cellularity approach was fully detailed by Gates (1986) as an attempt to standardise, in "cells", the electronics hardware. This focus was seen as the key upgrade/adaptability issue for electronics (which change so rapidly), rather than addressing the rapid replacement of the weapon delivery components of the combat suite. That it did not get adopted by the R.N. was seen at the time as due to a lack of whole ship vision across naval procurement, combined with the, then appropriate, combatant design (the Type 23) being conceived as a single role (Towed Array ASW) vessel design with short life, margin-less and cost (UPC) capped. Thus the ship project refused to have a likely small increase in volume for Cellularity capability, which was, in part, hard to demonstrate with the (pre-architectural) numeric based concept design practice (Andrews, 2003). This then resulted in a potential through life opportunity being lost, fleet wide.

It might be argued that the most successful implementation of container based mission bays has been the US Navy's LCS classes of fast mono-hull and trimaran combatants, which is certainly true numerically, since vessels are still being added to the US Fleet. However the programme has been somewhat disastrous (see article in Warship Technology, 2016) and in this instance it could be argued has meant clearly demonstrating the effectiveness of modular payload adaptability has been missed. This is largely due to the design conflict arising from the (questionable) very high combatant speed (50knots "desirable") demanded by the requirements owner (OPNAV).

However, a paper by Courts and Broadbent (2018) to last year's Warship Conference, summarising a series of studies by BAE Systems and UK MoD (DSTL), shows that the attraction of "Adaptability, Flexibility and Modularity" remains. That paper gives a comprehensive set of definitions and structured aspects where these three characteristics (or Styles, see Andrews (2017)) can be addressed for their advantages and disadvantages in being adopted in future naval combatants. It is not appropriate to repeat their paper's detailed presentation, which is very coherent and sets out to clarify what can be seem to be an ill-thought through bandwagon. That the paper also addresses the "optimisation" of Fleet Structure, as part of the "modularity" debate, indicates the need to make such "Style" choices beyond individual ship programmes (in contrast to the Type 23 above). This further, but rightly, complicates early stage ship design (ESSD) decision-making (Andrews, 2018b).

4. THE ISSUE OF MARGINS IN THE DESIGN FOR ADAPTABILITY

In the above examples of adaptable warship designs the term "margins" has been used to characterise, to a significant degree, the provision for adaptability. The consequences of deliberately omitting margins for Adaptability through life, in order to meet the political constraint of minimum initial cost, can be seen in the example of the Type 23 as already described. While the US Navy example of the CG 47 in Section 2.6 shows that exploiting an initially adaptable design can only be taken so far, as the issue is not simply accommodating predictable weight growth and the rise in the ship's centre of mass. It is the case that it is also necessary to retain a sensible reserve of space and ensure other ship features have sufficient margins to support the demands of adaptability. While it may not be the whole story on adaptability, the issue of what is meant by margins in naval ship design needs to be appreciated as a major part of the issue. What initially follows is essentially the UK practice as the approach adopted by the U.S. Navy uses a somewhat different terminology (Gale, 1975).

The concept of margins in warship design is quite sophisticated as there are essentially three distinct sets of margins. Only one of these is primarily concerned with adaptability, that one known in UK practice as Board Margin(s), which reflects that the level of Board Margin to be provided in a RN ship has traditionally been in the gift of the Admiralty or Navy Board. Whether that is strictly still the case, following a series of organisational and process changes, culminating in the Smart Procurement Initiative (UK MoD, 1999), is debatable; however the point being that these margins are in the gift of the operational requirement owner (i.e. the naval staff), rather than the designer, at least as far as intent, if not implementation. Of the other two categories, the first, Design and Build (or Contingency) Margins are required through out the design and build process to allow for uncertainty in the design, as it is being developed through to its definition on completion and these should all be consumed once the ship's fabric and equipment is manufactured and installed. These margins, essentially in weight, space and location, provide sensible budgets for the various sub-system designers to work within. These margins reflect uncertainty and without them the overall design balance of the ship would not be maintained, and redesign would be necessary late in the design and, furthermore, the conclusions of the earlier cost-capability trade-offs invalidated.

The other category, being that of Growth Margins, is often confused with the Board Margins as it might be seen to imply that it is something to be traded off to enhance the initial capability. This is a profound misapprehension, as Growth does not refer to growth in capability, which might be defrayed for enhancing current capability, but rather allows for unplanned growth in a warship's weight. Such growth has been seen to occur in warships over their operational lives, at a clear, measurable and reasonably constant manner. Without incorporating these margins in the design, the ship will only meet its stability, structural strength and powering standards and requirements on build. This would have grave consequences, either for ship safety or the operational freedom of the ship early in its life. It would almost certainly preclude any meaningful consideration of upgrading the operational capability - rather the opposite. However there is a clear need for the requirements owner and the ship designer to understand what is a sensible design point in the intended life of the individual ship beyond which at least some degradation in performance, if not in inherent ship safety, is acceptable. If this is not done then effectively the ship will spend the bulk of its existence over designed. The normal way out of this bind has been to plan on a mid life modernisation, one element of which is restore the ship back to safety standards for the remainder of its planned life. In this way Growth Margins are only required until this modernisation, however approaches such as Progressive Upkeep, if adopted, will mean this should not be so likely to be easily accomplished. Finally, it can be seen that while Growth Margins do not provide for Adaptability, they are intimately linked to that provision and that there needs to be coherency in the application of both sets of margins.

So the Board Margins provide for installing future equipment and, very importantly, provide for that equipment's likely impact on the overall ship design. Given the future equipment will demand space (often high in the ship), add weight, require additional crew, ship services and even enhancements to ship's structure, then the overall impact may be far greater than the direct physical attributes of any, currently speculative, new capability. It can be seen from this that those highly adaptable designs already discussed, by virtue of their generous design features, have been able to effectively exploit their Board Margin provision. Conversely, ships like the Type 42 just to maintain vital standards, required considerable additional refit work to install new capabilities. Furthermore incorporation of vital upgrades has been extremely expensive due to the knock on effects, which have required major work to restore standards, usually at the loss of some performance in, typically, speed, endurance and seakeeping. So Board Margins, while seen as essentially weight and stability (and possibly space) provision for future capability enhancement, should be combined with a sensible margin in other critical features of the design, such as local structural features and ship services. If this is not done, this may result in too much of the Board Margin provision being absorbed with the wider consequences of installing a reduced level of new equipment. This will then mean that the extent of capability upgrade possible could be noticeably reduced.

The US Navy's (NAVSEA) practice on ship design margins differs, at least in terminology, from the UK's approach, above, and it is worth pointing out some differences if only to understand Table 1 above, which has been re-produced here as it provides such a clear and salutary example. US Navy practice was summarised several decades ago by Gale (1975) and that paper is still pertinent. Thus "Future Growth Margins" are seen as specifically for Adaptability and then split into (OPNAV specified) FGM (equivalent to UK Board Margin) and "service life growth" (equivalent to UK Growth Margin/Design Point). The danger of overly focusing on the weight and KG implications of adaptability is shown by the consequences of the mid-life upgrade of the USS Franklin Delano Roosevelt (FDR). Blisters were added to the hull to "restore GM" after a major refit to accommodate the latest aircraft. Unfortunately, the additional GM was excessive and the carrier was too stiff in roll motion to then operate the aircraft. The issue with GM (often not realised by non-naval architects) is that it has to be sufficient for stability but still relatively small to avoid roll resonance. Given it is also the difference of two large numbers then a relatively small percentage error in either can lead to too little or (in the FDR case where the Chief Engineer of the Navy was sacked) too much GM.

However getting design provision for adaptability is not just weight, KG and space, it is also having sufficient margins in each of the distributed ship service systems (DS3). Any DS3 provision for adaptability can be hard to allocate early in the design, yet beyond that additional demands to the DS3 are difficult to meet and can have extensive "knock-on" implications. Current research, sponsored by US Navy Office of Naval Research (ONR) at several universities, including the author's research team at UCL (Brefort et al, 2017), are looking at better sizing and routing of DS3 in ESSD. Electrical load growth to accommodate adaptability demands is another tricky design call, although this may be eased in the next generation of naval combatants given fuller all-electric power schemes (IFEP) with more flexible installations.

5. DESIGN FOR ADAPTABILITY

Considering wider ship design related issues relevant to adaptability, beyond the crucial issue of margins, there are both generic combat system issues and others that might be better considered as more directly ship design related. However, any sharp division into combat and ship "halves" of the whole system of systems constituting a highly integrated modern warship runs the danger of falling into the bad design mind-set of "payload" and "platform", which was roundly dismissed in the introductory remarks in the author's paper at last year's Warship Conference (Andrews, 2018a).

5.1 COMBAT SYSTEM ASPECTS

Many current weapon systems have some degree of modularity/adaptability through being made to interface with ship elements, such as silos. Future systems such as Directed Energy Weapons (DEW) are likely to be seen as more integrated with regard to those components beyond the delivery elements of ordinance in silos, as shown by Figure 4 taken from the UCL DEW investigation (Andrews et al, 2011). The fit shown in Figure 4 is for two DEW units (a Free Electron Laser and a Long range Rail Gun) showing how they would be traditional installed – but could be readily modularised (and include sufficient adjacent space provision for future upgrades). DEW ship fits are likely to have a clear synergy with Integrated Full Electric Power (IFEP) installations already being proposed to "eliminate the tyranny of the shaft line" (Andrews et al, 2004). The ability the latter gives to placing prime movers and other major power distribution components in other than the traditional machinery spaces also opens up the possibility of more readily modularising such items. This is because they can be placed higher in the vessel with ease of such modules being more readily replaced, as is seen with the QEC carriers (Eddison & Groom, 1997) in comparison the complex removal lift arrangements for deeply located diesel generators in the previous INVINCIBLE Class (Honnor & Andrews, 1980).

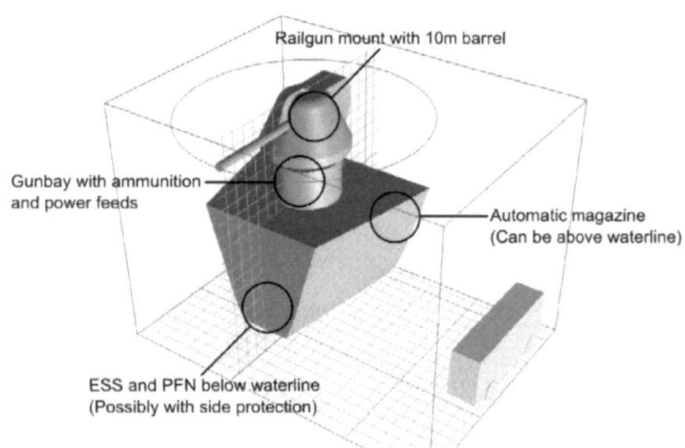

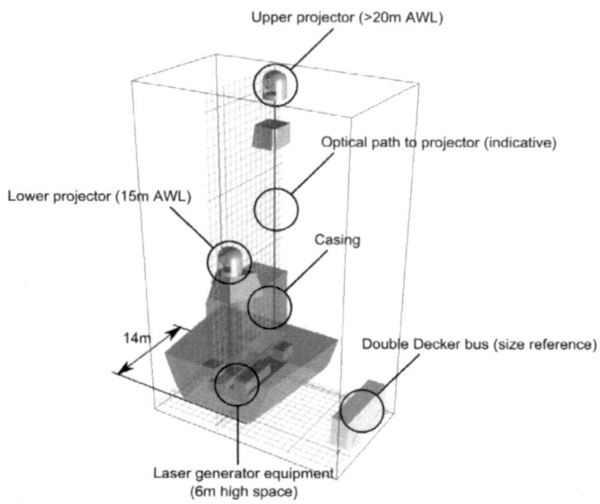

Figure 4: Generic weapon system layouts for a DEW Cruiser design (Andrews et al, 2011)

It has long been recognised that even the helicopter version of organic aviation, while giving naval combatants state of the art ASW location and prosecution capabilities, imposes a major ship fit demand on relatively small vessels, especially as the helicopters get bigger and the supporting features also seem to expand – see the Merlin implications on the Type 23 Frigates. (An excellent outline of such ship fit demands was provided by Farrell et al. (1971) where two SeaKing ASW Helicopters were incorporated into the design of the Tribal Class (DDH 280) for the R.C.N. Those authors concluded that such a fit dominated the aft half of that ship design.) If accommodating one large helicopter is combined with several unmanned air vehicles (UAVs), which are likely to be more regularly up-dated than manned rotary aircraft, then the case for providing adaptability to the aviation support features on-board would seem to be very strong. Whether this means modularising every part of the aviation outfit, given the ARAPAHO experience outlined in Section 3, seems questionable.

Another capability to be considered here, even if not solely arising from direct combat system related up-grades, is that of the adaptability of personnel provision onboard. While the ARAPAHO experience would seem to rule out "containerising" personnel accommodation demands, on the basis of excessive spatial demands for the relative small naval combatant, making the ship more adaptable to take likely increases in personnel provision as part of adaptability needs to be addressed early in an "adaptable" ship design. While there are on-going efforts to reduce warship's complements through automation (such as the step change for the US Navy with the Zumwalt Class) the earlier examples of both the USN FFG7 escorts and the R.N. Type 23 Frigates (see Andrews, 2018a) of arbitrary accommodation limits, mandated "to keep down initial procurement costs", run counter to achieving a coherent measure of adaptability.

5.2 SHIP RELATED ISSUES

Court and Broadbent's (2018) consideration of the related issues of Adaptability, Flexibility and Modularity led them to conclude that the intentions behind these related initiatives were best described by six classic (naval) ship design transversals (or "ilities"), which they call "Desirable Characteristics" and are included in this author's macro and major Design Style aspects (Andrews 2017). Courts and Broadbent's "six" are Producibility, Operability, Supportability, Flexibility, Re-configurability and Adaptability. This author would like to just briefly address (from a Design for Adaptability consideration) Producibility plus two other "Style" or transversal aspects, that of Survivability and the related issue of Stealth. The latter doesn't easily become an "ility" and, incidentally, spans the (artificial) divide of combat system and "the rest" of the warship.

Stealth might be readily seen as contributing to Survivability but raises specific design issues with regard to achieving Adaptability. The easier set of stealth aspects to address, in this regard, are those associated with above water signatures. In particular, radar cross section signature levels, vital to ensuring anti-missile decoys are effective, require the above water external structural shaping avoids right angle corners and planes to be some 7 degrees out of orthogonality. This is an issue if an adaptable solution makes recourse to containerisation or non-RCS compliant modules are exposed to missiles using radar seekers.

However a more fundamental ship design implication arises with underwater signatures, the key to ASW (Hamson, 2016). If a naval combatant is to be a truly effective ASW vessel then it must be designed from the very beginning with minimising underwater signature as a key design driver. Achieving this requires a comprehensive approach, not just for the hull form and appendages (hydrodynamic flow, particularly at key sonar operating speeds and not just at maximum speed for which the hydrodynamics of warships are usually "optimised") and the propulsors in the hull's wake (again specific speeds are crucial) but also **any** significant machinery on-board (particularly those low in the hull with noise paths) may need remedial attention up to and including "double acoustic mounting". Although most of the final signature profile will need addressing right through to the detailed design stages, without the overall "Style" being right from design commencement to reflect this major design driver, a first rate ASW capability cannot be achieved. From an adaptability stance any up-grade or new ASW sensor is hard to predict and, even if the design is generous, likely to be hard to incorporate.

Survivability (for a through consideration see, Piperakis et al, (2014)), particularly the Vulnerability component, could be said to be the traditional feature that distinguishes the warship from the merchant vessel, yet interestingly is not in Courts and Broadbent's (2018) six Characteristics. What Vulnerability means for an adaptable design can be considered in two parts:

Firstly, the level of resistance to major damage designed into the ship must be maintained as the features incorporated for adaptability are (rightly) exploited through life. Thus margins provided for weight and KG would include provision for still achieving the "design intent" with regard to Survivability. This can lead to an interesting challenge as "excessive" initial margins on weight, space and KG might mean, for a considerable part of its life, that the ship is (apparently) "over designed". It can also raise the question as to what design condition is the ship being designed to meet – that before its "adaptability" margins have been used or later in life once it has been up-graded. What price then the measured mile on acceptance or initial manoeuvring trials?

Secondly, if the design (as UK combatants should) incorporates the very effective but demanding feature of NBCD zoning (incorporated firstly post the Falklands Campaign in the Type 23, see (Manley, 2001)), this can present some difficulties if adaptability is achieved through adopting a large mission bay (as in the Type 26). A potentially large mission bay, such as that shown in Figure 5 (from Kouriampalis, (2019)), that might straddle two zones could invalidate substantial (double bulkheads) zone boundaries, or result in USS ZUMWALT like size of combatant that very few navies could contemplate (see concluding remarks).

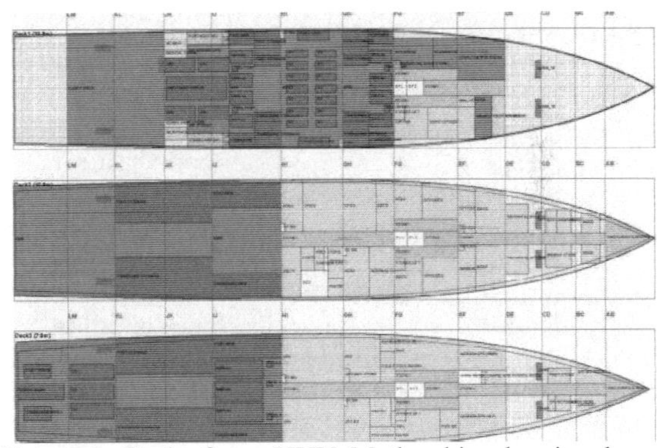

Figure 5: Study for a UXV Mothership showing how large Mission Bays can drive up ship size (Kouriampalis, 2019)

Producibility has been a warship design Style issue that has arisen from concerns with affordability (Keane & Fireman, (1993), Andrews et al, (2005)), however, a focus on initial procurement cost (UPC in UK terms – see the conclusion of Andrews (2018a) for a critique of UPC as a measure of affordability) can be seen as inconsistent with keeping the overall fleet expenditure down. Building into a class of new naval vessels the virtues of Adaptability, as propounded by the theme of this conference, is likely to be in direct conflict with Producibility. This is in part due to the latter currently being measured, in ESSD, by largely weight based (and historically derived) costs, which see the “smallest” ship solution as best meeting UPC “minimisation”. This obviously penalises incorporating such measures as adaptability margins and features requiring inherently “generous” designs, so revealing little regard for the subtleties of reducing through Life Costs.

In concluding consideration of Design for Adaptability, it is relevant that Courts and Broadbent (2018) explore the topic by looking at “Fleet Structure” rather than at the level of a single ship class that has been a particular policy mandated to be applied in isolation by each warship IPT Leader. Courts and Broadbent present Benefit vs. Cost Pareto Fronts for some five parameters, but still use UPC for the cost measure. A concluding result from these initial studies seem to be that a sub-set of capabilities in multi-purpose vessels is “more cost-effective” (albeit with the above cost limitations) than smaller role specific or even larger “fully flexible” vessels. However, this is said to apply only if multi-purpose means general escort and other (minor) roles, rather than prime Force capabilities, such as ASW and AAD. An issue not explored by this MoD sponsored study is whether major “Ship Style” design options (Andrews, 2018b) might realise the benefit if they were to be configured for Adaptability, even in the case of the top end combatants that Courts and Broadbent seem to rule out. One noticeable option in this regard is considered in Section 6.

6. TRIMARAN – THE ULTIMATE ADAPTABLE COMBATANT

The 2001 paper (Andrews, 2001) presented a review of likely hull configurations as part of a survey of trends in warship design. It is not intended to repeat that review here but rather to point out that some eight configurations were compared with the mono-hull for nine aspects which had previously been considered to be the basis, in general, for comparing different hull forms for a naval combatant (LRS, 1988). Clearly the choice of aspects is debateable and, in any particular new design, those most relevant to its emerging operational concept and acquisition constraints will actually determine the assessment of the relative merits, rather than necessarily those shown in the table, which are of general applicability. It is noticeable that the nine (LRS) aspects did not explicitly include that of Adaptability or, indeed, Modularity. Although adaptability could be said to be implied in the final aspect in Table 2 of through life costs, it has been considered from early on under the S^5 item of Style by the author (Brown & Andrews, 1980) and recently expanded upon, (Andrews, 2017). A more explicit set of aspects could be the High Level Characteristics (derived under the UK's Smart Procurement Initiative), which were adopted for the RN's Future Surface Combatant (Andrews, 2000). The FSC, as the precursor to the Type 26, was seen to have seven High Level Characteristics, three of which were directly missions related, Warfighting, Power Projection and Utility, while the others were considered to be enabling, namely, Interoperability, Sustainability, Survivability and, most significantly for this paper's theme, Adaptability. Thus Table 2 has the additional aspect of Adaptability in addition to the 2001 version.

What might seem remarkable from the, admittedly qualitative, exercise comparing the various hull configurations summarised in Table 2, is the very impressive score of the Trimaran configuration, even compared to the mono-hull, which has been the invariable choice for all, except for the most operational specific naval vessels. This latter fact has been due to the mono-hull being an inherently mission flexible hull form able to undertake most naval roles with more than reasonable facility. It is argued once more that the Trimaran has the potential to replace the mono-hull as the most effective configuration for the warship of the future, not just because of the assessment suggested by Table 2, but specifically because it provides, more readily and with less design impact, those features that are required for Adaptability. Unlike the other configurations compared in Table 2, the ocean going Trimaran combatant is essentially a relatively minor departure from the mono-hull (Pattison & Zhang, 1995). It has many features in common with the mono-hull and can draw on the proven technology applicable to mono-hull warships, unlike most of the more advanced forms in the comparison. That the only extant Trimaran combatant, the US Navy's LCS variant, is more technologically radical than the original UK concept (realised in the two-thirds scale demonstrator for the Trimaran option of the FSC, R.V. TRITON (RINA, 2000)) more due to the high top speed sought for the LCS role than due to its trimaran configuration.

There are considered to be two aspects in the overall design nature of a mono-hull warship that noticeably restrict its ability with respect to the need to provide an adaptable design. They are the ability to cope with maintaining adequate stability with subsequent technology insertion and providing sufficient space on the upper works and main deck(s), especially amidships. These aspects are necessary to achieve an adaptable design but there are also other aspects, which have been mentioned earlier. The origin of the Trimaran came from a desire to separate the traditional dichotomy in the mono-hulled warship, that is to achieve the necessary stability, particularly with the growth in top-weight with the ever increasing demand for higher and larger sensors (such as the Sampson Radar so significant in the Type 45), while requiring a long fine hydrodynamic form to minimise installed power at high speeds. The Trimaran achieves the latter by providing a very fine main hull and the former by the relatively small side hulls, which by virtue of their wide separation, provide sufficient intact and damage stability required for very little additional immersed displacement. Furthermore, the Trimaran can do this more effectively than the mono-hull as the latter is limited to that moment gain which is set at its design

Table 2: Qualitative Assessment of Different Hull Forms (amended from Andrews, 2001)

Aspects \ Ship Types	Mono-hull	Cata-maran	ACV	SES	Hydro-foil	SWATH	HY-SWAS	WIG	Trimaran
Speed, Power and Endurance	Good	Good[1]	V Good[2]	Good	V Good[2]	Good	Good	V Good	Good
Space and layout	Good	Good	Ave	Good	Poor	V Good	Poor	Poor	V Good
Structural design and weight	V Good	Ave	Poor	Poor	V Poor	Ave	Poor	V Poor	Good
Stability	Good	Good	Good	Good	Good[3]	Good	Good	Poor	V Good
Manoeuvrability	Good	Ave	Poor	Good	Good	Ave	Good	Poor	Good
Noise, Radar & Magnetic Signature	Good	Ave	Good	Good	Good	V Good	Good	Good	V Good
Weapon placement & effectiveness	Good	Ave	Ave	Ave	Poor	Good	Ave	Poor	V Good
Construction costs and build time	V Good	High	V High	High	V High	Good	High	V High	Good
Through life costs	Good	Ave	V High	High	V High	Ave	High	V High	V Good
Adaptability	Good	Good	Good	Good	Poor	Good	Good	Poor	**V Good**

Notes.
1. But bad in deep ocean seaway
2. Very fast but limited to hull borne (slow) in seaway and endurance poor (fuel weight)
3. Very good hull borne but foil borne degraded by wave effects in deep ocean

freeze, whereas the Trimaran can if necessary, accommodate any reasonable level of rise in top-weight. This can be met by relatively minor modifications to the side hull separation. It amounts to effectively a tuneable feature through out the ship's life - a characteristic providing noticeable KG margins for adaptability only when required.

However, adaptability also requires internal space precisely where in a mono-hull "real estate" is most competitive – high in the hull amidships, where generosity in space provision is sensible if an adaptable design is to be achieved. Thus the Trimaran's cross deck structure, linking the three hulls, provides the ideal configuration for an adaptable ship. In sizing a new Trimaran design, one of the major design decisions concerns the wet deck clearance, from still water to the underside of the cross structure (Andrews & Zhang, 1995). This gives the designer essentially excess volume in the ship, relative to the mono-hull equivalent, in the cross deck structure, at amidships and high in the ship. With the weather deck above the cross structure being considerably wider in the middle third of the Trimaran it can much more readily accommodate large hangars located with the flight deck close to the ship's pitch centre, a feature very hard to achieve in a mono-hull. Additionally, as Table 2 indicates, the Trimaran also scores well for other issues identified as crucial in achieving an adaptable warship, readily meeting the survivability, supportability and producibility features better than its mono-hull equivalent.

With these advantages one might question (even with the (slightly odd) example of the US Navy's high speed Trimaran LCS Class) why the Trimaran has not been seriously pursued in recent R.N. combatants, given the R&D lead UCL and the UK MoD established in the late 1990s. The case for the Type 45 being trimaran configured might just be countered by its parallel design development in that period, although the author (reinforced by his personal involvement in both programmes) considers the lack of consideration more due to MoD "handing over" Design Authority for the Type 45 to its Prime Contractor (Gates, 2005) as being symptomatic of procurement policy overriding design clarity (Andrews, 2018a). The case for the Type 26 to be trimaran configured seems even stronger with the Trimaran option being studied as part of the FSC concept explorations (Andrews, 2000) and, arguably, **the** design driver being Adaptability. Again there was a degree to which the procurement policy decided not to explore this potentially advantageous novel solution post FSC, where the mind-set appeared to be overly risk adverse. This might be perpetuated in the current competition for the Type 31e "Exportable" light frigate, despite the above arguments that if you want Adaptability (say for export customers with differing requirements) a highly adaptable design should be a winning one. If a Trimaran combatant can give multirole adaptable solutions for relatively low risk, why have we not been able to move on from operationally "optimised", initial cost constrained and highly tuned design solutions to now design and build truly adaptable warships for an uncertain future?

7. CONCLUSIONS ON MULTI-ROLE AND ADAPTABILITY

It would seem, in meeting the needs of multi-role capability, that adopting adaptability characteristics is likely to be the way forward, but this is far from straight forward since the main issues and conflicts this paper has identified need to be addressed from the very earliest in the combatant design process. The author has recently argued (Andrews, 2018b) that the ESSD for complex vessels, such as naval combatants, is sophisticated and must not be rushed through "to get on to design proper" (i.e. post the concept phase). Some of the issues raised in this paper reinforce the need for sophistication in deciding whether to adopt a typical major Design Style issue – that of Adaptability.

This sophistication means exploring, in a given new naval vessel programme, the implication of the desired level of Adaptability on the various consistent capabilities (such as Courts and Broadbent's "six Desirable Characteristics" plus other "transversals", such as have been discussed in this paper). Secondly, the exploration should include a comprehensive consideration of the solution space (Andrews, 2018b) as this could better reveal the means of meeting the perceived needs for the new programme. An indication of thinking "out of the box" has been given by consideration of the adaptability of the trimaran configuration in Section 6.

Nevertheless, pursuing the full capable multi-role warship solution can run the risk of ending with a single affordable "Starship Enterprise" (Kirkpatrick & Pugh, 1982) so there is a need to ensure coherence between solution and requirement in producing a balanced design. In this instance one with sufficient Adaptability, taking cognisance of the historical examples of First Rate options and avoiding spurious "optimisation", such as the Type 23 as a Cold War GRIUK Gap concept being then used as a general ocean going combatant, would better ensure adaptable designs can be pursued. That this is best done by also considering novel solutions, which might be battle winning over safe (non-risky) designs, ought to be part of a properly conducted ESSD process (Andrews, 2018b).

8. ACKNOWLEDGEMENT

The author acknowledges the production and use of various design studies by members of the UCL Design Research team, however the views expressed in the paper are solely those of the author.

9. REFERENCES

1. ANDREWS, D.J. (2000), *Perspectives on the Future Surface Combatant,* Sea Change 2000, Jane's Conference, Southampton, May 2000.

2. ANDREWS, D.J. (2001), Adaptability – the Key to Modern Warship Design' RINA Warship 2001, London, June 2001.

3. ANDREWS, D.J. (2003), *A Creative Approach to Ship Architecture*, RINA Trans IJME, Vol.145, Sept 2003.

4. ANDREWS, D.J. (2011), *Marine Requirements Elucidation and the Nature of Preliminary Ship Design*, IJME Vol. 153 Part A1 2011.

5. ANDREWS, D.J. (2017), *Choosing the Style of a new design: The Key Ship Design Decision,* Trans RINA Vol 159, Part A4, IJME, Oct-Dec 2017.

6. ANDREWS, D.J. (2018a), *The Conflict between Ship Design and Procurement Policies*, RINA Warships 2018, London, Sept 2018.

7. ANDREWS, D.J. (2018b), *The Sophistication of Early Stage Design for Complex Vessels*, Trans RINA, Special Edition, IJME, Vol 160, 2018.

8. ANDREWS, D.J., BUCKNALL, R., PAWLING, R., GREIG, A. & MCDONALD, T. (2011), *The Impact of Integrated Electric Weapons on Future Warship design Using Conventional and Unconventional Hullform*, Conference on the Ship as a Weapon, IMarEST, RNC Greenwich, Sept 2011.

9. ANDREWS, D.J., BURGER, D. & ZHANG, J.W. (2005), *Design for Production using the Design Building Block Approach*, IJME, Vol. 147, RINA, 2005.

10. ANDREWS, D.J., GREIG, A.R. & PAWLING, R., (2004), *The Implications of an All Electric Ship Approach on the Configuration of a Warship*, INEC 2004, Amsterdam, March 2004.

11. ANDREWS, D.J. & ZHANG, J-W. (1995), *Considerations in the Design of a Trimaran Frigate*, Int. Symp. High Speed Vessels, RINA, Nov. 1995.

12. BAKER, R. (1956), *Habitability in Ships of the Royal Canadian Navy*, Trans SNAME, 1956.

13. BETTS, C.V. & HARRIS, P. (1984), *Conversion of Merchant Ships for Military Operations*, RINA Warships Conference, London, June 1984.

14. BREFORT, D. et al. (2017), *An Architectural Framework for Distributed Naval Ship Systems*, Ocean Engineering, Vol. 147, 2017.

15. BROOME, G.W. & FRIEDMAN, N. (1992), *Conjecture: Future Surface Combatants*, SNAME/ASNE Naval Ship Design Symposium, Arlington, VA, Feb 1992.

16. BROWN, D.K. (1983), *A Century of Naval Construction*, Conway Maritime Press, London, 1983.

17. BROWN, D.K. & ANDREWS, D.J. (1980), *The Design of Cheap Warships*, Proc International Naval Tech Expo '80, Rotterdam, June 1980.

18. COLLINS, J.T. (1975), U. S. S. SPURANCE (DD963) *Class Designed for Change - The Payoff is Now*, U S Naval Engineers Journal, April 1975.

19. COURTS, M. & BROADBENT, C. (2018), *The Implications of Adaptability, Flexibility and Modularity for Royal Navy Warship Design*, RINA Warship Conference, London, Sept 2018.

20. EDDISON, J.F.P. & GROOM, J.P. (1997), *Innovation in the CV(F) – An Aircraft Carrier for the 21th Century*, Warship 97 RINA International Symposium on Air Power At Sea, London, June 1997.

21. FARRELL, K. et al. (1971), *The DDH 280 Class Design*, SNAME Eastern Canadian Section, Feb 1971.

22. FIGG, K. (1994), *The Evolution of Future, More Affordable Warships*, MILTECH, 6/94.

23. GALE, P.A. (1975), *Margins in Naval Surface Ship Design*, USNEJ, April 1975.

24. GATES, P.J. (1986), *Cellularity, an advanced weapon electronics integration technique*, Trans RINA, Vol. 128, 1986.

25. HAMSON, K.G. (2016), *A New approach to managing the Acoustic Signature Risk in Warship Design*, RINA Warship Conference, Bath, UK, June 2016.

26. HOCKBERGER, W.A. (1993), *Cost and Operational Effectiveness Analysis (COEA) in Naval Ship Design*, SNAME Chesapeake Section, May 1993.

27. HONNOR, A.F. & ANDREWS, D.J. (1982), *HMS INVINCIBLE - The First of a New Genus of Aircraft Carrying Ships*, Trans RINA Vol. 124, 1982.

28. KEANE, R.G. (1982), *Discussion to The Design of Variable Payload Ships by Broome et al USNEJ*, June 1982.

29. KEANE, R. & FIREMAN, H. (1993), *Producibility in the Naval Ship Design Process: A progress Report*, JSP, Vol.9 No.4, Nov 1993.

30. KELSO, F. (1992), *From the Sea: Preparing the Naval Service for the 21st Century*, US Naval Institute Proceedings, November 1992.

31. KNIGHT, S. (2009), *The Design of HMS QUEEN ELISABETH and HMS PRINCE OF WALES*, RINA 2009 Warship Conference, London, June 2009.

32. KORIAMPLAIS, N. (2019), *Applying QT and Architecturally Oriented ESSD to the Concept of a Vessel Deploying a Fleet of Uninhabited Vehicles*, PhD Thesis, UCL, 2019.

33. LEOPOLD, R. & REUTER, W. (1970), *Three Winning Designs – DD 963, LHA and FDL*, Trans SNAME, 1971.

34. LLYOD'S REGISTER, (1988), *Warship Hull Design Enquiry*,

35. MANLEY, D. (2001), *Procuring for Survivability, RINA Warship 2001 Conference*, London, June 2001.

36. PATTISON, D.R. & ZHANG, J. (1995), *Trimaran Ships*, Trans RINA, Vol. 137, 1995.

37. PIPERAKIS, A.S., ANDREWS, D.J. & PAWLING, R.J. (2014), *A comprehensive approach to survivability assessment in naval ship concept design*, Int. J. Maritime Engineering, Vol.156/A4.

38. PURVIS, M.K. (1974), *Post War RN Frigate and Guided Missile Destroyer Design 1944-1969*, Trans. RINA Vol. 96, 1974.

39. RINA International Conference (2000), *R.V.TRITON*, Solent Conference, Southampton, April 2000.

40. SPERO, J.R. et al. (1971), *A Philosophy of Naval Ship Design and Construction*, USNEJ, Oct 1971.

41. THOMAS, T.R. & EASTON, M.S. (1992), *The Type 23 Duke Class Frigate*, Trans RINA Vol. 134, 1992.

42. UK MOD, (1998), *The Strategic Defence Review, Directorate of Defence Policy*, HMSO, July 1998.

43. UK MoD (1999), *Smart Procurement – The Acquisition Handbook*, SPI Team, April 1999.

44. WARSHIP TECHNOLOGY (2016), *SASC leaders take aim at restructured LCS programme*, March

45. WATKINS, J.D. (1982), *Letter to the Chairman of the House of Representatives Committee on Armed Services on Committee's Staff paper on USS TICONDEROGA*, 23 August 1982.

9. AUTHORS BIOGRAPHY

David Andrews was given a new Chair in Engineering Design at UCL in 2000 when he retired from the UK MoD as a Director Surface Ships. He led the design integration of aircraft and weapons fits on the INVINCIBLE Class Carriers, was senior naval architect in the Forward Design Group before promotion to Warship Project Manager for the Amphibious Replacement Programme and a significant portion of the in-service fleet. He was subsequently Head of Preliminary Design in Future Projects responsible for initial design and procurement studies, including those on the Future Surface Combatant, CVF and the WAVE Class tankers. In 1993-98 he was appointed MoD Professor of Naval Architecture at UCL where he directed the MSc in Naval Architecture and led UCL's innovative research into Trimaran ship design. In 2001 he set up a new Design Research Centre in the Department of Mechanical Engineering, which is focused on computer aided ship design (CASD) and the design of complex systems. He is a Fellow of RINA, a Vice President and Member of Council. In 2000 was elected a Fellow of the Royal Academy of Engineering and is currently the International Chair for the tri-annual series of the International Marine Design Conferences.

WHEN IS THE ADAPTABLE OR MULTI-ROLE SHIP THE RIGHT ANSWER?

C E Moss and **A C Kimber**, BMT UK

SUMMARY

The rise of the multi-role ship, often envisaged with modular bolt on capabilities, has been the stable of conferences for most of this decade. Similarly, adaptability is an increasingly common theme in presenting new designs and concepts. However, the question for the naval architect is "do these represent the right solution for a given individual project" or has the concept been overstated as a vision of the future and used as the template which is not quite the panacea that it promises. After several decades involved in a wide range of projects, the authors offer their thoughts on the opportunities and realism of multi-role ships and adaptability. The paper explores some of the lessons learnt in delivering multi-role and modular designs. The underlying themes that are fuelling the interest in multi-role ships along with the typical roles considered are also discussed.

1. INTRODUCTION

Adaptable and multi-role ships appear to offer benefits. Of course the perceived wisdom is that a multi-role ship delivers more capabilities cost effectively than many ships and adaptability mitigates the unknown nature of future operations such as differing types of warfare, environment, threats or technology.

But why is a multi-role or adaptable ship design automatically perceived as the answer rather than an adaptable *Fleet*? In fact engineering logic could conclude that two different tasks are best performed by two specialised ships rather than one. Or it may be better to replace obsolete ships with ones designed for new systems at more regular intervals rather than construct ships with large margins on the basis that they have correctly predicted the characteristics of future systems and operations. Many commercial ships are designed for one purpose and are replaced rather than upgraded.

Much seems to be driven by the politics of defining a number of hulls within a navy or the difficultly of initiating replacement programmes, which drives towards a limited number of ships with extended operational lives. Hence the natural answer of an adaptable or multi-role ship, irrespective of its cost effectiveness; it is born of political necessity. In the past we perceive that threats were more stable, or at least the RN was large enough to dedicate ships to each perceived threat. Today, the limited number of ships and the wide ranging / changing tasks dedicates ships of multiple capabilities and (at least) perceived adaptability.

In fact it is the nature of naval ships (auxiliaries or warships) that they are inherently designed for multiple roles. However, beyond today's ships, many navies strive for the one size fits all warship or auxiliary that does everything rather than a ship that does just a number of closely related roles.

2. WHAT DO WE MEAN BY ADAPTABLE AND MULTI-ROLE?

Initially we would like to clarify what we mean by an adaptable and a multi-role ship as at times these terms have been used interchangeably, however they are different types of ship. For the purpose of this paper the following definitions will be used:

- Adaptability – is defined as to be able to adapt to new conditions or be able to modify to a new use or purpose [1]; this implies an adaptable ship is one that can be upgraded with new equipment for its existing role or be changed to a new role during its service life. The perception here is that by designing in adaptability the ship is easier to change, by adding margins or modularising equipment to allow easier exchange;

- Multi-role – is defined as being capable of performing several roles [1]; this implies a ship that is designed to be able to fulfil a number of specific / defined roles throughout its design life. The ship may be able to perform roles concurrently or be able to be reconfigured without changing the platform itself to switch between roles.

It should be noted that the above are not mutually exclusive; a ship may be multi-role but not adaptable, it may be adaptable for future upgrades within a single role or it may be both multi-role and adaptable.

3. WHAT DOES SUCCESS LOOK LIKE?

The aim of the naval designer is to provide a successful ship to the customer. The success of a naval ship can be measured a number of ways including; cost, value for money, survivability and operational effectiveness to name a few and is not necessarily the same for all classes of vessel. As suggested above; reduced hull numbers, rapid changes in technology and uncertainty in operational requirements leads to multi-role and adaptability being seen as fundamental requirements. For example, in a recent survey [2], when asked "What will be the most important component of the future support fleet", design modularity was returned by far as the dominate response, at 39% (availability scoring next at 22%). The same survey yielded statements such as "multi-role is critical for maximum return on the investments made".

So, considering some historic examples does it follow that multi-role and adaptable ships imply successful designs?

Whilst many of the "modular" adaptable ships entering service or on the drawing board have yet to prove themselves, the lessons of applying good margins in design can be seen. In [3] the case of the Spruance class is a good example of this. Designed with future modernisation in mind, these ships were designed with generous margins and appeared large for their initial capability, drawing some criticism. However, these ships were able to undergo not only significant upgrades resulting in an increase in displacement from 8,400tons for an unmodified ship to 9,400tones for a heavily modified ship, but the hull became the parent for the later and more capable CG47 class (although in this case the subsequent CG47 now lacked the margins which made the original parent so adaptable) [4]. Hence, building in future adaptability can be successful in a design, although this success is realised only at the point the margin is used and if it isn't it may be viewed only as a compromise to the primary purpose.

Additionally a naval ship may be seen as being successful if it is designed for a specific role but through its design it is able to be adapted to perform alternative roles; examples of this include the Invincible (CVS) Class and potentially the QEC aircraft carriers. The Invincible design can certainly lay claim to being a successful design as they operated successfully as ASW helicopter platforms, aircraft carriers, LPH's and well beyond the original design brief (referred to as a "victim of her own success" in the [5] due to the extensive changes incurred at refits). The QEC has learned the lesson of the Invincible Class and has adaptability and multi-role capability inherent from the design stage, which hopefully will remove the need for the large refits. Although very early in their service life, the QEC physical size and inherent flexibility means that although being designed as a carrier strike platform they have the potential to be successful ships for other operations. In both cases these represent large, capital ships of high programme cost, intrinsically linked to a changing technology (aircraft) with a shorter life than the ship.

Consider also the example of WWII aircraft carriers. The UK designed its carriers with armoured flight decks and hangars whilst the USN adopted wooden flight decks. The wisdom of WWII suggests that the RN carriers were more survivable, better able to cope with damage and able to more quickly recover operations than their US counterparts. It could be concluded that in this respect they were a more successful design of capital ship. However, consider also the situation in the 1950's when both navies began to adopt bigger jet aircraft; the USN converted many of its WWII era carriers, the RN struggled as the design was hard and costly to modify, chiefly because of the armoured decks [6]. So we could conclude that the RN Fleet Carrier of WWII was a successful design because it was very good in the role and environment for which it was designed whilst the USN Carrier was successful because of its adaptability to operate future aircraft types. Success is therefore relative with respect to when you look at it.

A good example of multi-role versus single role are the Type 12 and Type 14 frigates. Broadly of the same era and both with an ASW role using the same system (the Limbo mortar), the Type 12 frigates offered a more general capability around the ASW role whilst the Type 14 frigates focused solely on delivering their ASW role with little additional capability provided. Table 1 illustrates the difference in size of these two platforms. History would seem to favour the Type 12, which was later converted to carry helicopters and also developed into the highly successful Leander Class frigates, whilst the Type 14 frigates proved limited and paid off before the Type 12's. However, this is perhaps unfair, as the Type 12 suited the prolonged cold war era whilst the cheaper Type 14 may have been more successful had the RN needed to rapidly increase numbers (they were intended to be mass produced if needed). This also illustrates a key point; multi-role ships offer the flexibility needed over long periods of peacetime where tasks change, but the added size and complexity drives cost and build time. Single role platforms can boost numbers due to lower cost and can have a quicker procurement cycle.

	Type 12	**Type 14**
Length	112.7	94.5
Beam	12.5	10.1
Displacement	2560	1456
ASW Weapons	2 x Limbo Mortars	2 x Limbo Mortars
Other Weapons	1 twin 4.5inch	2 x 40mm
Sensors	Hull sonars (Types 174, 170, 162 Navigation Radar (Type 975), Search radar (Type 293)	Hull sonars (Types 174, 170, 162 Navigation and search radars (Type 975, 978)
Machinery, Speed	Twin Shaft, 31kts	Single shaft, 27kts
Crew	225	140
Number in Class (RN)	12	14
Construction Period	1952 - 1955	1952 - 1958
First of Class Cost / Build	£1.434M / 1246 days	£3.081M / 1379 days
Average Class Cost / Build	£1.826M / 1249 days	£3.329M / 1295 days

Table 1: Type 12 and Type 14 Comparative Data

One obvious disadvantage of multi-role vessels is that the number of vessels available to undertake concurrent activities is reduced. If two, or more, roles are required concurrently or at different locations and cannot be undertaken a single vessel then this leads to operation precedence decisions with potentially some roles not being undertaken.

A multi-role ship is usually designed to undertake a defined set of roles and if used for these, or a number of these, throughout its operational life it would be seen as a success. However, if it is only used for one of these roles throughout its operational life, even if it successfully undertakes this role, is the vessel a success? There is an argument that it is not, as a simpler and cheaper ship designed for that specific role could have been procured instead to deliver the same capability.

Equally, where cost is constrained for a ship, it is potentially more beneficial to focus on one role. From [7] Brown and Andrews suggest "There is a distinction which must be drawn between limitation of capability, which implies full performance in a limited role, and reduction of performance, which could include a multi-role ship inferior in every aspect. The authors have shown that a single role is the only viable approach to a cheap warship but, like most philosophies, this must not be carried to extremes." The conclusion is that when faced with constraints on cost or size it is better for a ship to do fewer roles well rather than more but poorly.

For a multi-role vessel to be successful, when adopting a modular solution, the whole package needs to be considered, designed and procured as a single system. Assuming that modules will be developed and procured later risks the concept never being fully realised and unrealistic assumptions on the ability to re-role may lead to adoption of permanent configurations and as such an unsuccessful design.

Defining what is required through to disposal of a ship can often span 40 and up to 50 years for the more complex ship types. As such the ability to predict what will be required in the future and defining the amount of adaptability in the design is paramount to being able to deliver it (as depicted in Figure 1). However as this is very difficult to do in a rapidly evolving world (both with respect to political situations and technology), flexibility in one's ships becomes very attractive (adaptability with regard to keeping up with technology and multi-role capability with regard to role uncertainty), especially with reduced fleet sizes.

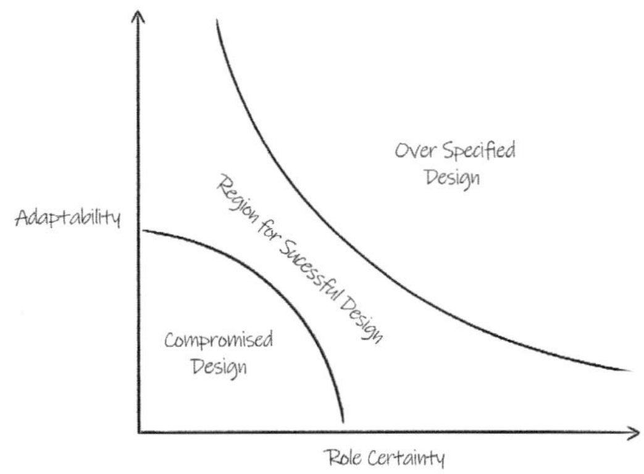

Figure 1: Adaptability vs. Role Certainty Graph

As can be seen by the above discussion there is no single answer to what a successful ship is. One that performs to its specification, within budget and delivery schedule? Or to exceed its specification or even change roles through life?

Success is judged on the context of the project. If the vessel is expected to have a long career, perhaps in a peaceful period, then its ability to adapt over time to changing politics and technologies may prove its success. If the navy has limited hull numbers then multi-role ships offer the promise of more capabilities, whilst a larger navy may value the ability of an individual ship to deliver against its core role and provide flexibility at a fleet level. Where cost, time or size is constrained, then designer should focus on the primary requirement and deliver this well, then see what else is achievable whilst not compromising the primary role. If all roles are equally weighted then they will be in competition with each other resulting in all being compromised and the vessel may not be able to do anything well. So as designers, we should understand why a customer may seek a multi-role and adaptable solution to ensure it presents the best solution and the intent is coherently applied throughout the design.

4. DESIGN DRIVERS

Whilst the definitions of adaptable and multi-role ships are different the key design impacts of both are similar and include:

- Increased ship size to accommodate more mission equipment or margins;
- Increased accommodation to allow for more operators, more roles or future growth in capability;
- Increased systems and / or system capacities to accommodate different / more mission equipment or margins;
- Performance characteristics (speed, signature, survivability, etc.) are required to match the

requirement of the most demanding role or anticipated adaption.

There are a range of factors that drive ship designs including standards, speed, stability, size of crew, roles, etc. However, when considering adaptability and multi-role ships there are a number of additional things that need to be considered and the importance of some of the more traditional factors may increase. Key factors are listed below, addressing adaptability and multi-role ships separately but noting that the adaptability discussion are also pertinent to multi-role ships.

4.1 MULTI-ROLE

- Ability to conduct roles concurrently or via reconfiguration of the vessel;
- Known roles / number of roles that the vessel is to be able to do;
- Conflicts between roles (space, manpower, power, etc…);
- Time to re-role:
    - Short term: To be undertaken at sea / organically on transit requiring storage space and reconfigurable spaces;
    - Medium term: Return to shore and swap modules (modules need to be available and as such procured with the ship and /or procured within foreseeable sight of need);
    - Long term: Return to base for significant re-rolling (refit activity). Design to minimise disruption on majority of the ship. This may be considered as adaptable as well as multi-role.
- Trade between size constraints and space required for multi-role systems, which may constrain concurrency;
- Physical interfaces and services regarding the integration of modular equipment into the platform;
- Software interfaces between modular equipment and the platform systems.

In terms of designing a multi-role ship, the first key driver is to understand whether the roles are concurrent or achieved by the exchange of systems, equipment and / or personnel. Designing for concurrent roles is addressed through providing sufficient space, equipment and personnel to deliver all the roles. This simply increases ship size to accommodate all the functions, de-conflicting functions through size / location. Provided there is no conflict in any of the required characteristics, this should be straightforward. RFA FORT VICTORIA [8] is a good example, it can replenish both fuel and stores, and also has a large hangar and multisport flight deck to support ASW operations and aviation training. It resulted in a large ship, with significant accommodation which was costly compared to other contemporary auxiliaries, but has high utility beyond the basic replenishment function.

When developing multi-role ships we see a general increase in vessel size for any generic type of vessel, this being more pronounced for the smaller classes of ship, e.g. frigates as can be seen with the current UK Type 26 frigates which are comparable in size with the Type 45 destroyers. Additional issues may also arise from "political" decisions to constrain ship size without trading capability, which can lead to complex and difficult technical issues in incorporating all of the capability (often with differing space, power and weight requirements) into a single constrained ship. This can lead to cost escalation which in turn can lead to issues in procuring the required numbers of ships for the fleet, as a whole, to undertake what is required of it.

The more challenging approach is to design a multi-role ship which attempts to constrain costs and / or size through re-role and adaptability. The difficultly here is that it would be a rare case where the roles require the same weight, space, systems or personnel. This can be further compounded when different roles are required concurrently but were not designed to be undertaken concurrently. Hence, as a designer it is important to recognise the compromises being made, which role has supremacy and the operational practicalities and compromises of the proposed solution.

The Norwegian Logistic Support Vessel (Figure 2) is presented elsewhere in this conference but represents a ship that can perform a number of roles, but through adjustments in the personnel embarked and some re-rolling of spaces. Ensuring the ship was operationally viable in its many roles was a key feature for the Norwegian customer. This should include consideration of how the ship will be re-roled and the infrastructure support required, how the personnel on-board are impacted and how supplementary personnel are provided, trained etc.

Figure 2: Norwegian LSV (HMNoS Maud)

The drivers behind multi-role support ships was further explored in reference [9]. Whilst illustrating that multi-role capabilities are already delivered by ships such as the RFA FORT VICTORIA, this paper explores how increasing the number and diversity of roles can add complexity, cost and compromise to a design. This can occur because whilst auxiliary ships may be considered as a common group of ships, they in fact can have very different characteristics that do not sit comfortably in an single ship hull design (for example a replenishment tanker is a weight driven design with a fuller hull, a logistic support ship is volume driven where the hull is shaped to maintain a suitable draught).

The search here is for synergy in the platforms characteristics that allow it to perform two or more roles. As an example, the BMT VENARI®-85 design [10] was created to offer a future MCM vessel capable of utilising off-board vehicles. As such a vessel may reasonably expect to find its self in the area where the mine threat is "grey", therefore the vessel is afforded some signature management. To minimise the chance of activating acoustic mines, one feature is a noise reduced propulsion fit. This has synergy with ASW platforms, therefore one option subsequently developed is for the platform to deliver an ASW sensor role through an alternative mission fit including a modularised towed sensor deployed over the stern (Figure 3). By adding single and multi-beam echo sounders and replacing the MCM boats with hydrographic boats, it also can perform hydrographic missions. This platform can now perform three missions – MCM, hydrographic and ASW detection – based on a coherent set of platform characteristics such noise reduction.

Figure 3: BMT VENARI®-85 - MCM to Multi-role Ship

4.2 ADAPTABLE

- The intent of incorporating a defined level of adaptability e.g. upgrade of systems, ability to change to an unplanned role or a combination of the two;
- The size of margins to be applied to address future updates with technology advances e. g. weapons and sensors;
- Trade between size constraints and space required for adaptability, e.g. dock infrastructure, etc.;
- Physical interfaces and services regarding the integration of new equipment into the existing platform;
- Software interfaces between new equipment and the existing platform systems;
- Understanding the potential technology changes and cycles.

An adaptable ship shares many of the same considerations as a multi-role ship. However, future adaptability may not always be well defined, and may simply be an aspiration to keep pace with future innovations, e.g. today the call is for ships to operate unmanned systems but often with no specific vehicle in mind. Often this will translate into margins applied to the design. Again, the designer must consider which margins, are they coherent and if they impact the overall ship viability. Often forgotten is the need to ensure the ship is not compromised at delivery by having applied significant margins which are not yet consumed, resulting in a ship that may initially have too lighter draught, poor seaworthiness (rectified perhaps by ballast) to name a couple. However, harder to fix are over specifying systems such as electrical generation, HVAC or other systems where the resultant capacity is too great for the initial load making them highly inefficient. This relates back to the comments regarding successful designs, i.e. too much adaptability via margins may lead to an un-successful ship when delivered and a reputation that may never be shaken. This suggests a more iterative approach between the customer and designer to "balance" the roles and adaptability against ship operability is necessary.

The use of modular systems is one approach to minimise the impact of adaptability (and is also applicable to multi-role ships) by standardising the interface between ship and capability. Several projects have made this central to their approach and today many navies see the standard Twenty Foot Unit (TEU) container as the building block. A good description and appraisal of approaches to modularisation is provided at [11]. This paper illustrates the complex nature of trying to draw together the many themes under the banner of modular and adaptable systems, including the immaturity of the procurement processes and the lack of coherent specifications for modular systems.

A key choice here is whether to adopt a centralised area for embarking these modules (Type 26, Littoral Combat Ship or Finland's Squadron 2020) or to spread the modules to specific locations (more akin to the Danish Stan Flex approach). Large mission spaces can offer a benign environment compared to open decks and allow multiple adjacent modules; however they demand volume, handling systems and can be challenging to incorporate into a design (e.g. structural integration and continuity, access routes, survivability of the modules); some of this is further explored in [12]. Alternatively, module spaces or adaptable margins could be distributed to specific locations representing the ability to enhance or incorporate future upgrades to capabilities with specific characteristics. This is potentially easier to integrate within a design but more constraining on the possible variations. An example would be the Stan Flex approach adopted in the Iver Huitfeldt design, which has module spaces arranged in a number of locations (Figure 4). A similar approach was adopted by the BMT VENATOR®-110 design [13].

Figure 4 - Iver Huitfeldt – Flexibility [12]

The physical space for modules is only one dimension and in order to fully realise this approach it is important that the impact of each configuration and the potential for concurrency is flowed into the ship's systems. Hence, it needs to be reflected into the weight estimates, electrical, chilled water and other system load charts and scheme of complement. This is where adaptable and multi-role ships can become challenging to the designer, i.e. maintaining coherency across many design artefacts produced by the design team requires a watchful eye by the design manager and clear design instructions to the team. A system engineering derived approach and a suitable shared design environment are essential. Some of the wider considerations for adaptability and flexibility in a design are covered in the [14].

5. SPECIFYING ADAPTABILITY

As adaptability / flexibility is ethereal, not directly measurable at acceptance, how do we specify and accept it? Can it be as simple as weight, space and other margins or is it more complex than that?

In the past, when the designer and owner were within the same organisation they could apply good practise and lessons learned directly to a design as it evolved. Today, most navies procure ships against a set of requirements with industry developing the solution to meet those (often contractual) requirements. Hence, the requirements are set before the design is commenced, demanding that there is a clear characterisation of the required adaptability, expressed in terms that can be accepted as part of contract completion. This has many advantages, but in this context it does require adaptability to be robustly defined, i.e. how adaptable, in what way and how is it demonstrated? Although it is possible to explore margins and adaptability in concept and illustrative designs, the system engineering approach illustrated by the V diagram (Figure 5) can drive out flexibility. It's a process that sequentially links need to requirement to solution to acceptance, but a design's solution space is not necessary linear (as explored in [15]) and this approach limits the designer's scope to create a balanced design and exploit opportunities. This is also considered in [16], which discusses the "middle out approach" as an alternative with a more parallel design and requirements activity. This is not an easily resolved issue, again it focuses on successfully achieving the requirements or developing a successful design?

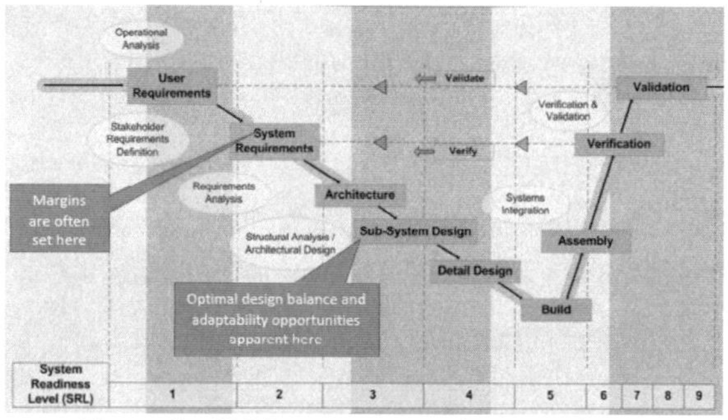

Figure 5 – V Diagram approach versus design optimisation

In summary, there are three broad approaches to adaptability that may be adopted:

- Build big: Just add extra space, buoyancy and system margins. History says building bigger will add future flexibility and versatility (D. K. Brown concluded that with a few exceptions "good ships are versatile as completed and adaptable to new roles. In general, such ships were generously sized as built" [17]. As balancing the margins is design specific this concept works well where there is a close relationship between designer and customer, allowing exploration of the best size. But as discussed above it requires a leap of faith that the extra space is cost effective and this may not survive the procurement process which may require tight bounding of the requirement.

- Build modular: Pack functions into more easily replaceable units. This approach needs agreed interfaces, unit sizes, etc. which have been proposed locally (Stan Flex in Denmark or the US LCS Mission Packages for example) but have yet to be released at a more international level, although work is ongoing within NATO [11]. This is also based on the assumption that the mission equipment fits in a standardised box. Most naval ships, warships or auxiliaries, now include some capability to host modules, but the fully modular ship has proven elusive.

- Build in specific upgrade provisions: Use generic assumptions to define what "future" may look like and allow for specifically allocated space, weight and system margins where they are perceived as being needed. This approach is adopted in the Type 31 programme. It does of course require a good understanding of how future systems may develop to create the right sort of provisions and does not fully address the unknown unknowns. However, it may at least create sufficient potential to mitigate even unexpected changes.

All three of the above processes have their advantages and drawbacks but it is clear that a close relationship between designer and owner is a route to achieving the best results. Indeed incentivising the designer to make the design adaptable would be a good approach, if the commercial mechanisms and acceptance criteria could be defined.

6. DESIGN AND BULD CAPABILITY

There is a link between design and build demand, the choice between a multi-role or multi-class fleet, and procurement cycle versus adaptability.

Where a country wishes to retain a national design and build capability, then the fleet size will determine ship building sustainability, whereas the number of types of vessel will determine design sustainability. A key choice for a smaller navy is whether to procure multi-role or adaptable ships which may not provide sufficient industrial loading to sustain national design and build capability, or to adopt shorter life and simpler vessels at reduced cost but increased frequency. Larger navies are better able to sustain build capacity, however the design capacity is still dependant on the need to develop new designs (as highlighted in [18]).

A smaller navy may have to think differently with regard to its procurement strategy for its fleet as the smaller the number of ships it has the harder it is to maintain a consistent delivery programme / demand on its ship yards and design capability. Cost and hull number constraints may lead such navies to adopt more multi-role vessels to provide them with the range of capability that they require / wish to have. This can lead to a bust and boom profile for in country ship yards unable to sustain / win orders from other countries. Many navies have been faced with such decisions and have had to accept that they will need to buy in designs from another country and possibility have ships built abroad as well; others are facing this issue currently. Such out of country procurements usually result in a longer operational life for the platform to minimise the frequency of large offshore expenditure.

7. STANDARDS

Ship design rules, standards and guidance have developed over time for the various classes of naval ships with their specific considerations. However, in moving to multi-role ships the divisions between these become less clear and therefore a greater degree of tailoring is required to develop a comprehensive rule set that is also coherent and safe. This is further complicated, over that of commercial vessels, by the diverse and evolving operational use of Naval ships which standards do not necessarily keep up with.

As there are no clear standards for multi-role vessels each programme needs to expend significant effort to develop its own specific suite of tailored documents. Care has to be taken in partial application / tailoring of standards as a mix and match approach is difficult and can result in an unsafe solution.

8. CONCLUSION

The answer to "When is the Adaptable or Multi-Role Ship the Right Answer?" is not a simple one and one that is dependent on a number of factors including; budget, size of fleet, role(s) of the ship, operational life of the ship and a country's desire to maintain its own design and build capability.

The basic tenants of adaptable or multi-role are founded in seeking value for money and through life cost effectiveness. But as discussed in this paper, understanding the detail of how these are implemented and the resultant complexity can lead to convoluted discussions on what the intent is, how it is to be specified and the best approaches of achieving a successful design. They have a transversal impact on the design and success is only really understood after some years of operation.

There is no single factor that determines if an adaptable or multi-role ship is the right answer to a particular ship procurement cycle. However, if the right questions are asked and robustly considered early in the procurement cycle it can be possible to procure a ship that can be truly successful by avoiding an over, or under, specified vessel because it is the fashionable or easy thing to do.

Among the questions the designer should be asking are:

- How adaptability and multi-role requirements are delivered in light of the whole fleet mix in order to achieve value for money?
- Are the roles well defined and realistic looking forward to the potential operational life of the vessel?
- Are the wider implications of re-role and adaption understood, e.g. personnel, training, infrastructure?
- Is it understood what the primary role is and how it is impacted by the other assigned roles?
- Is the concurrency of roles and the time scale to re-role clearly defined and implications understood?
- Is the ship actually operable in all roles with regard to personnel, ships system loadings and synergy of performance characteristics understood?

- How is adaptability to be incorporated, e.g. margins, modules, mission bays, etc.?
- Is the required operational life of vessel understood along with robust assumptions on the potential for changes, i.e. the sizing of margins?
- Is the variance of performance with and without margins consumed addressed?
- Are the required standards defined and suitable?
- How will the procurement process impact the ability to specify and deliver a successful design?
- Will the procurement process constrain / drive the relationship between the designer and owner, and will this compromise the design?
- Is it understood how adaptability will be specified and accepted?

The success of a design is always judged with hind sight and as such at the outset it is necessary to understand as much as possible about the operating requirement and political landscape in which the vessel will operate throughout its life.

9. REFERENCES

1. Oxford Dictionaries | English. (2019). English Dictionary, Thesaurus, & grammar help | Oxford Dictionaries. [online] Available at: https://en.oxforddictionaries.com [Accessed 24 May 2019].
2. 'Support Ships Market Overview', *Published by IQPC*, 2019.
3. SIMS P., LIN J., 'Common Studies Necessary to Support Early Stage Alternative Evaluations', *American Society of Naval Engineers Naval Engineers Journal 2013.*
4. KEANE R. G., TIBBITTS B. F., 'Fallacy of Using a Parent Design: "The design is mature"', *Society of Naval Architects and Marine Engineers 2015.*
5. KERR Captain R. G., 'Invincibility – changes in HMS Invincible during refit', *Journal of Naval Engineering 1988.*
6. BROWN D. K., MOORE G., 'Rebuilding the Royal Navy: British Warship Design Since 1945', *published by Chatham Publishing 2004.*
7. ANDREWS D. J., BROWN D. K., 'Cheap Warships Are Not Simple', *American Society of Naval Architects and Marine Engineers 1982 (reproduced by JNE 1983).*
8. MORRIS, G, MULLIGANM R. D., RAYFIELD B, COOKE J, 'AOR - The Auxiliary Oiler Replenishment', *Journal of Naval Engineering 1988.*
9. KIMBER, A. C., 'Joint Support Ships - Trumping the Jack of All Trades', *MAST Europe 2009.*
10. AITKEN A. J., 'The Next Generation MCM Platform – Not Yet Full Autonomy', *Royal Institution of Naval Architects 'Warship 2018'*.
11. DOERRY N. H., 'Institutionalizing Modular Adaptable Ship Technologies', *Journal of ship Production and Design, August 2014.*
12. KIMBER A. C., GILES W., 'Minor Warship Roles - How technology is leading to a new vessel type', *Pacific International Maritime Conference 2009.*
13. KIMBER A. C., 'Designing for the Gap: The space between the OPV and the Frigate', *International Ship Design and Naval Engineering Congress (ISDNEC) COLOMBIMAR 2015.*
14. JOHNSON J. WAKELING I., 'Flexible Design as an Acquisition Opportunity', *International Ship Design and Naval Engineering Congress (ISDNEC) COLOMBIMAR 2017.*
15. ANDREWS D., PERCIVAL V., PAWLING R., 'Just How Valid is the Ship Design Spiral Given the Existence of Cliffs and Plateaux?', *International Marine Design Conference 2012.*
16. KIMBER A. C., THORNE B., 'How to Square the Circle for the Next Generation of Patrol Ship Designs', *Royal Institution of Naval Architects Warship 2013.*
17. BROWN D. K., 'History as a Design Tool', *Royal Institution of Naval Architects 1992.*
18. SCHANK, SAVITZ, MUNSON, PERKINSON, MCGEE, SOLLINGER, 'Designing Adaptable Ships – Modularity and Flexibility in future Ship Designs', *RAND Study 2016.*

10. AUTHORS BIOGRAPHY

Charles Moss holds the position of Chief Naval Architect (Surface Ship Structures) at BMT. He has over 23 years of experience of working on naval vessel and has significant design experience on a number of programmes including the QEC aircraft carriers, Tide Class Tankers, Norwegian LSV and the UK's Fleet Solid Support programme.

Andy Kimber holds the position of Chief Naval Architect for Surface Ship Design at BMT. He has 30 years of experience as a Naval Architect in the defence industry, covering concept studies, design delivery and in-service support. He has held positions in the QEC design team, Tide Class (leading the design) and one of the Type 31 bid teams in the UK and has been involved in projects for the US, New Zealand, Australia, Canada, Norway and Finland amongst others. He has also been involved in projects considering fleet mix balance, options for capability delivery and procurement approaches.

THE IMPORTANCE OF ON-BOARD INFRASTRUCTURE TO ENABLE FUTURE VESSEL ENHANCEMENTS AT BUILD AND THROUGH LIFE

S Willmore, BMT, UK
C Burnside, E Sutton, BMT, UK

SUMMARY

Increasingly the rate of technological change means that platform configurations and capabilities are evolving. Often equipment installed during build was not available or accurately specified to be incorporated early in the design cycle. Developing engineering approaches and good stakeholder engagement is needed to ensure that design choices consider future likely or potential developments by providing a degree of flexibility without expanding budgets to handle the uncertainty and risk to the programme. Platforms are now updated and upgraded regularly through their life; often platforms within a class vary in the capabilities they hold due to the rapid change in technology and the mission profiles for the platforms. In an information centric battlefield with the craft of information warfare swiftly becoming increasingly prevalent, connectivity is fundamental to mission success and so future topside layouts need to effectively support this and consider antenna positioning as a key enabler to optimising this. Below decks the ability to setup and re-configure networks both functionally and physically is key to allow for low cost future updates avoiding the need for design iterations and lengthy installations of new networks. Providing secure, high capacity links between other local units, organic assets and ashore will be essential if developments in domains such as autonomous systems are to be fully exploited to further advance overall platform effectiveness.

GLOSARY

ADAWS – Action Data Automated Weapon System
AESA -Active Electronically Scanned Array
ATM – Asynchronous Transfer Method
CEC – Co-operative Engagement Capability
CMS – Combat Management System
COTS – Commercially Off the Shelf
CVS – Small Conventional Aircraft Carrier
ELINT / COMINT / SIGINT – Electronic / Communications / Signals Intelligence
EMCON – Emissions Control
EOIR – Electro-Optic and Infra-Red
GFE – Government Furnished Equipment
II – Installation Inspection
IR – Infra-Red
LDEW – Laser Directed Energy Weapon
LEO – Low Earth Orbit
LPD – Landing Platform Dock
LPH – Landing Platform Helicopter
MAS – Maritime Autonomous Systems
MEO – Medium Earth Orbit
NWHT – Naval Weapons Harbour Trial
NWST – Naval Weapons Sea Trial
RadHaz – Radiation Hazard
RESM – Radar Electronic Support Measure
STW – Set to Work
T42 – Type 42 Destroyer
UAV – Unmanned Aerial Vehicles
X25 – X25 Network Standard
5G – 5th Generation Mobile Telephone

1. INTRODUCTION

Advances in technology mean that commercially available technology particularly in relation to computing, networking and communications is now driving future equipment procurement in defence. For warships and submarines the challenges are stark. Whereas in the past Combat Systems typically would be designed ahead of the ship or submarine design and sometimes would cover multiple platforms (ADAWS CMS on T42 B1-B3, CVS, LPH and LPD), now the lifecycle of the underpinning commercial technology requires obsolescence management to replace part or all of these systems within the lifetime of the build phase for a large class let alone the in-service life of these platforms. The designer and builder of these naval platforms often will not know the exact hardware being fitted to the first of class ship until well into the build and outfitting process, which creates a challenge, let alone know what the final ship in the build cycle will be fitted with.

Ensuring that platforms have the flexibility to undergo major reconfigurations of these networked and interconnected systems is a challenge.

On the horizon are new technologies such as 5G satellite communications, autonomous systems and digital twins which will pose the platform designer greater challenges as the need to integrate new equipment and operate with systems that may not be procured or indeed even defined for decades will be common place.

Modern navies will need to be able to rapidly update capabilities and add new functionality to platforms if they are to maintain a strategic advantage over their adversaries. Strategic or operational advantage will be eroded with naval platforms and their personnel being increasingly put at risk in conflict and hostile scenarios if no action is taken to adopt new technologies. Delivery of the "Fight capability will become increasingly unaffordable and financially inefficient due to existing procurement models and system designs. Moving to a model with a greater use of software defined capabilities running on distributed and shared infrastructure enabled

by the use of Commercially Off-The Shelf (COTS) technology. Supporting naval platforms will also benefit from greater use of data which will need to be supported by appropriate on-board infrastructure to provide connectivity to enable data to be analysed ashore.
In order to meet these challenges the design of future warships must consider this in the design, building and support of future naval platforms.

2. WHAT DO WE MEAN BY ON-BOARD INFRASTRUCTURE

For the purposes of this paper we are restricting the definition to the infrastructure that provides connectivity between individual equipment and systems as well as computing services that host or support a range of individual systems.

Below or between decks is where the majority of hardware is housed. In an increasingly networked environment, naval platforms often house a large number (20+ in some cases) of individual networks being used to link individual elements of a system together, or providing a wide area network linking lots of independent systems together. Some networks are wired and increasingly making use of commercial network technologies like Cat 5 Ethernet, fibre optics etc. as the carrying medium with a myriad of network standards in use such as ATM, X25 or IPv5.

Advances in computing power and a move towards cloud based or shared computing is now starting to allow equipment providers to share hardware and software elements. For the Royal Navy the Shared Infrastructure programme is migrating the surface fleet to an architecture where computing environments, networks, system services and consoles are able to be shared between equipment.

The proliferation of mobile communications devices and wireless networks are now being seen on naval platforms; from Infra-Red communications in engine rooms to Wi-Fi providing welfare services. The flexibility of wireless networking makes it an attractive option due to the range of commercially available options although the practicalities of transmission, being inside metal structures, and the requirement for multiple topside antennas as well as the security concerns around EMCON can limit its use.

Increasingly naval platforms operate as part of a wider information node where information is exchanged between units in the same locality or globally or with organisations ashore. Platforms are increasingly reliant on good connectivity to conduct operations and support routine business such as reporting defects or providing welfare links ashore.

These external links are established through communications aerials on the upper / top deck which are often vying for the available space. At some powers and frequencies these aerials when transmitting can cause interference or radio hazards (Radhaz) in the form of non-ionising radiation which affects personnel, explosives, fuels, other units and potentially the efficiency of on-board operations. Some aerials are isotropic and so radiate equally in all directions and some are directional and only work when they have a clear line of sight to the unit or receiver, introducing significant constraints when it comes to the matter of selecting a position on the platform topside. Managing mutual interference and managing the spectrum generated by multiple radio frequency sources requires careful management both physically (spatial separation) and frequency separation.

The upper deck also is home to the other sensors that are required for normal sailing or operations such as radars, EOIR sensors, RESM / ELINT, SIGINT and COMINT receivers and transmitters. Each sensor and directional communication aerial ideally needs to have the best coverage possible but as these systems proliferate, the challenge for the designer is finding a suitable location for each of these discrete functions that allow for the optimum installed performance and availability across these systems.

There is also considered to be redundant or latent capability within some systems that whilst the requirement was traded out inherent capability was left in the system. Improvements with infrastructure may enable the unlocking of this capability either by networking systems together or providing access to new sources of information.

3. HOW IS THE ROLE OF THE NAVAL PLATFORM CHANGING

Increasingly platforms are re-tasked into new roles mid-way through deployments. Due to reducing hull numbers in most navies this practice sees platforms operating outside of their original design intent conducting operations such as counter narcotics, counter piracy, strategic intelligence, maritime interdiction operations and humanitarian and disaster relief operations. Flexibility of naval personnel and ships is now required for all units, although most platforms were designed around a single high intensity war fighting role such as Area Air Defence or Anti-Submarine Warfare.

During a single deployment platforms will most likely operate as a singleton unit operating its own organic assets like helicopters, sea boats and maritime unmanned systems (MAS) such as UAVs. Whilst transiting to new areas of operation often units have to transit choke points like the Suez Canal which increases the risk of opportunistic attacks by terrorists; this increases the demand for sophisticated Force Protection measures or uplifts in capabilities such as Special Forces. Communication with these assets is increasingly important as more and more data is shared in real or near real time between units operating together.

Deployed platforms often work with other units and their organic assets, even if it is just in passing but also can work as part of a major task group conducting operations as part of a larger team. Increasingly interoperability between other elements of an armed service, national allies or other organisations (police and coast guard) highlight the challenges with interoperability particularly when exchanging data or secure voice messages.

The ability to be re-tasked, with units not identified at the point of leaving home waters at the beginning of an assignment, resulting in the need to, embark new systems whilst away from home waters is an everyday challenge for globally deploying navies.

4. THE ROLE OF THE SYSTEM INTEGRATOR IN SELECTING AND MANAGING INFRASTRUCTURE

The role of the System Integrator whether that be for the Combat System, Platform Systems as a whole, or for smaller systems such as the Navigation System, varies hugely between platform procurements. In this paper we have abstracted the role of the Integrator and presented only one particular set of relationships recognising that the role of the prime, design, System Integrator, contracting authority, operator and platform builder vary between procurements although the activities that need to be undertaken remain constant. For many navies the split between the designer / builder and supporter for the System Integrator may be constant throughout a platform's life or may transfer between organisations. It is also common for the System Integrator to employ Sub-System Integrators within the supply chain.

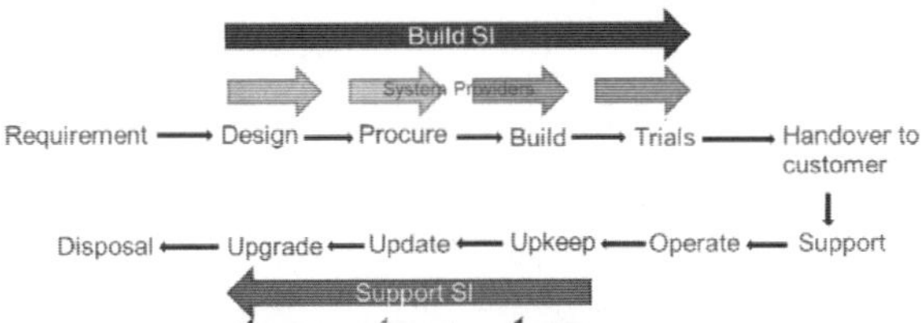

Figure 1 – The Simplified System Integrator Lifecycle

During the design and build phase of platform procurement the System Integrator is often required to translate the functional needs of the system, such as its capability requirements and interoperability needs, and their relationship to other systems, into physical things such as equipment and networks. Translating the customer need into a specification and then selecting appropriate supplier offerings can be a challenge often with innovation being traded out. For infrastructure, whilst novel solutions and increasingly complex network architectures are considered, redundant capacity is often traded out to make savings. Enabling infrastructure is frequently not specified by the customer and therefore misses this consideration as part of the standard v-lifecycle for maturing a platform design.

Increasingly platform procurements involve the inclusion of equipment provided as part of other government procurements or from legacy stockpiles. This Government Furnished Equipment (GFE) can represent a challenge for a System Integrator as equipment that may be unknown may need to be integrated both physically and functionally into a system but also be included in the design and supply chain processes.

The inclusion of GFE in a design could take several forms each with their own challenges:

- Provision of design information to inform the platform design and to underpin safety cases and security accreditation.
- Equipment delivery, often to an integration facility well ahead of platform installation
- Equipment support and upkeep whilst awaiting installation (for example some electronic equipment contains batteries that need changing if not connected to power supplies).
- Installation (or support to Ship Builder installation) and any formal Installation Inspection (II).
- STW (or support to Ship Builder STW) and presentation for any formal Naval Weapons Harbour Trial (NWHT).
- On-board systems integration (or support the Ship Builder or other appointed Authority for on-board system integration) and presentation for trial.
- Equipment support (defect repair and spares), between installation and ship handover, including provision of any Initial On-board Spares.
- Design support and/or configuration management, including any technology refreshes, between equipment delivery and ship handover.
- Trials support and/or presentation for any formal Naval Weapon Sea Trial (NWST).

The scope of supply will in large part depend on the complexity and maturity of the specific equipment. The more "commodity like" equipment will require minimum support (for example an echo sounder), whilst the more complex equipment (for example air defence systems) will require a full support package. Often the GFE supplied equipment are themselves subject to design evolution requiring the synchronisation of both the platform and the equipment design and production life cycles.

Apart from the installation and trials support that may be provided by a government provided agent, often the maturity of the equipment project will dictate how GFE is included in the design. For mature equipment that is unlikely to change it could be fully scoped and included in the design, whereas some equipment may be based on quickly evolving technology and so the equipment characteristics maybe too immature at the point the platform design needs to be set. Additionally some GFE may still be in early stages of planning or pre-procurement and so only the vaguest technical information is available.

GFE can be thought of being provided in 3 main ways:

- Full Fit – where the equipment technical information is used to mature the design and the actual equipment is delivered, installed, set-to-work and trialled as part of the platform build and acceptance programme.
- Installation Provision Made in Design – where equipment is maturing, classified or there are limited units a platform may have sufficient engineering work designed in and completed in build to allow the installation, set-to-work and trials to be completed with the minimal of additional engineering work by the government operator.
- Limited Provision Made – where very limited if any information is known provision could be left for space, weight, power and cooling margins. This represents a challenge for the platform designer as reaching agreement on suitable margins to use for these embryonic equipment characteristics is often unclear and sometimes over inflated margins are requested leading to over specified design features

For the System Integrator the on-board infrastructure to support GFE is often vague and may require the integration of legacy interfaces and network architectures. Technical information is often only provided at the Last Responsible Moment in the platform design process. Good technical governance and stakeholder engagement with the government customer is needed to ensure that decisions made on the available engineering information is fully understood and that risks and opportunities are managed.

Taking a more holistic approach to networks providing point-to-point connectivity and options for re-routing and re-configuring the physical network could make the use of GFE easier for the system integrator to respond to changes during build but also allow greater flexibility in-service.

Once in-service the System Integrator role is often split into sub-systems represented by suppliers or could be retained by the original Integrator. Importantly the role in-service is more to enable the future integration of GFE into the platforms existing infrastructure. Maintaining a view of available infrastructure and platform margins is essential to assure the continued safe operation of the platform.

With the move to more networked and interconnected systems the platform design at build needs to consider how it will be updated and upgraded through life. With platforms now lasting significantly longer than their on-board systems considering network physical architectures and how they can be extended, re-configured and replaced cost effectively should now be an important consideration of any platform procurement.

5. HOW WILL FUTURE TECHNOLOGIES IMPACT ON PLATFORM INFRASTRUCTURE

The pace of technological change for computers, communications, networking, autonomous systems and weapons technology is rapid and set to transform many aspects of maritime operations in the life time of the current and next generations of naval platforms.

5.1 COMPUTING

Changes in computer technology have already started a move from bespoke, single use computer platforms to thin or zero client consoles. Improvements in computing power have allowed equipment solutions to embrace simpler configurations with single servers replacing the role of dozens on desk stations.

The move to using virtual machines to run software in shared hardware but maintaining the appearance of standalone hardware is allowing the rationalisation of hardware in navies already. The move to virtual networking also enables greater access to data generated within equipment and system components in ways that allows greater exploitation. The exploitation of this data will transform the way operations are conducted and how platforms are supported. Virtualisation also enables the production of software only applications to run on a platform's infrastructure which could provide new capabilities rapidly to a platform by downloading and installing a new application. The move to software applications in this way reduces the often time consuming processes needed to ensure good engineering governance for physical changes to platforms.

Further out, the move to quantum computing is set to allow impossibly hard algorithms and computations to be completed nearly instantaneously. How this will be deployed in an operational context is unclear. Quantum Cloud Based services may enable platforms to send information ashore to be processed and streamed back to platforms. Or these systems could be used on-board to handle complex signal processing challenges. Quantum computing is still evolving but it is doing so in a way similar to Moore's law for semiconductors in the 20th century [1].

The drive for more complex computing on semi-conductor based computers is driving up the power requirements of these systems to power more powerful processors. The increase in power also develops more wild heat creation which needs to be managed as part of the infrastructure considerations.

5.2 COMMUNICATIONS & NETWORKING

Communications and networking to other platforms and ashore has been transformed by the introduction of geostationary satellite communications. In the next decade the introduction of CubeSat (small scale satellites) into Low Earth Orbit (LEO) and Medium Earth Orbit (MEO) is set to transform connectivity at sea again.

LEO based satellite communications will provide global coverage of high-speed data and communications including over the seas. Unlike geostationary systems which are trained to stare at particular areas these new constellations will be orbiting much closer to the Earth and so will appear to be moving rapidly overhead. To provide coverage over populated areas hundreds of satellites will be needed to provide coverage. For maritime platforms this will have the benefit of providing an affordable way of accessing high-speed internet connections anywhere at sea [2].

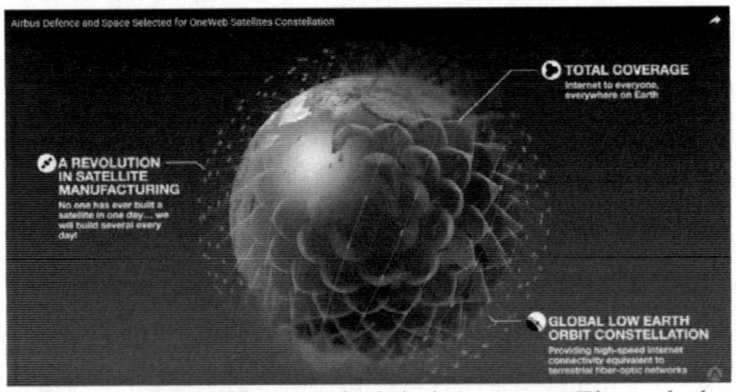

Figure 2 – Visualisation of Global Coverage Through the Use of LEO Satellite Communications [2]

Additionally, the introduction of the 5G standard for communications ashore will be supported by satellite based 5G services.

Changes to 5G infrastructure and in particular aerial design will make the use of Active Electronically Scanned Arrays (AESA) more common. For platforms wishing to make use of LEO satellite communications they will most likely need to use AESA planar or conformal planar arrays to track these fast moving satellites. Due to the apparent motion of the satellites, conventional, physically steered dishes will be unable to track a satellite and handover to the next one without breaks in transmission. AESA based technology allows for multiple links to be established concurrently at different points in the sky.

Communicating large amounts of data ashore will become more common place. With the move to Digital Twins supporting ship and submarine engineering as well as greater reliance on information warfare the need to share information will increase. Increasing connections from platform ashore from kilobits per second to hundreds of megabits and up to gigabits and beyond will need to become common practice. Platform design will need to embrace this ensuring that communications equipment is considered early in superstructure design to optimise performance and de-conflict with passive functions and other active transmitters alongside other transversal considerations such as signature management.

Networking with other units either for collective or force level engagements will require local communication networks to be established to along high bandwidths and low latencies. The United States Co-operative Engagement Capability (CEC) programme is an example where the communications links are as important as the weapons integration [3].

Similarly the use of autonomous systems will most likely require improvements in connectivity particularly for remotely operated or semi-autonomous platforms. In a similar way to CEC ensuring line-of-sight coverage between vehicles or using satellite communications will be needed.

5.3 AUTONOMOUS SYSTEMS

Autonomy and the use of Artificial Intelligence has the potential of reducing the risk to personnel by conducting hazardous operations such as mine clearance with unmanned platforms. The use of UAVs has been demonstrated for military and commercial uses with maritime equivalents currently under development and trials. Many off-board autonomous systems will require infrastructure improvements whether those are improvements in connectivity, in recharging or launch and recovery of the vehicles.

At sea line-of-sight communications is governed by the height of the transmit and receive aerials above sea level. Many autonomous vehicles are designed to have a low profile which means their aerials are close to sea level limiting their line-of-sight transmit range due to the curvature of the Earth.

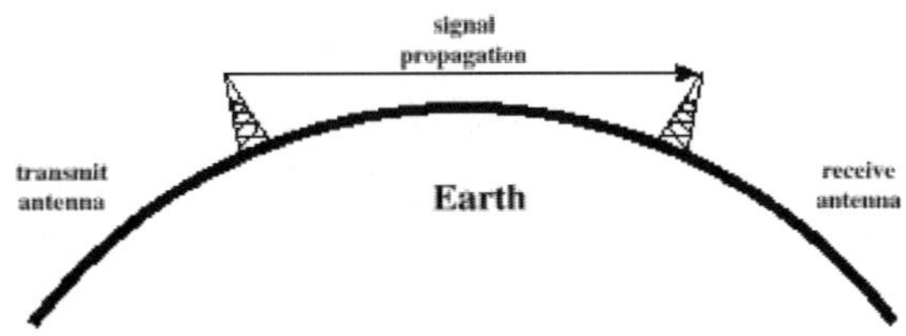

Figure 3 – Visualisation of the Line-of-Sight Path for Signal Propagation on a Curved Earth

An alternative is to use atmospheric propagation and transmit at lower frequencies but higher powers are required to "bounce" signals off the upper atmosphere. For autonomous platforms with limited power this is most likely not viable. Satellite communications is viable provided the cost of the bandwidth is low enough. For now considering antenna locations for communicating with off-board platforms will need to be considered during platform design.

5.4 FUTURE WEAPONS

The next generation of offensive air weapons that naval platforms will have to contend with are hypersonic missiles or anti-ship ballistic missiles. These missiles will be moving very fast and descending through the atmosphere on a ballistic trajectory towards naval platforms. Reducing reaction times, increasing the detection radius, using off-board sensors or other platforms will all be needed to help defend against the threat. For a singleton platform exchanging sensor information, threat information and targeting weapons between platforms will be essential to detect, classify, localise, identify and neutralise these threats [4].

Laser Directed Energy Weapons (LDEW) have the potential of replacing some weapon systems. The appeal is rapid fire rates and unlimited magazines. There are some drawbacks such as large power requirements and a need to cool a large heat source in the form of the laser itself. Whilst the visible director head is on the upper deck the laser cavity, heat sink etc. can be below decks and away from the director using fibre optics or waveguides to connect the laser source to the firing point. This allows some flexibility in the position of the directors, which may allow for greater coverage than legacy guns due to their need to be located close to magazines to allow for reload if not manual aiming. A further infrastructure consideration is the electrical architecture of the ship as the platform is going to need to be able to generate hundreds of kilowatts to megawatts of power to enable a laser with output powers over 20 kilowatts. Having a responsive electrical system that allows this capacity switching will be essential in future electrical system scaling and design [5].

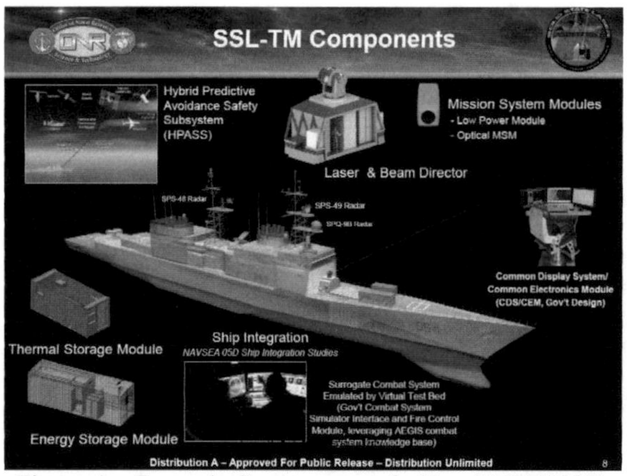

Figure 4 – Example Components of a Naval LDEW [5]

Rail guns make use of electromagnetic induction to accelerate projectiles rather than using propellants. The power needed for a shot or repeated shots is considerable requiring extensive power management. On-board electrical systems will need to consider this and provide energy storage to allow for electrical systems to charge up energy reserves for the rail gun to operate effectively.

6. ABOVE DECKS DESIGN CONSIDERATIONS

6.1 AERIAL COVERAGE

The location for aerials for sensors and communications equipment is important to ensure the required installed performance levels are met. Greater height increases the optical range by providing a greater line of sight coverage particularly when communicating with surface vessels. But increasing the height of aerials also increases the height of the vessel or adds more weight high up in the structure.

Some aerials are omnidirectional as they have isotopic radiation patterns. These aerials require less precise placement as they radiate in all directions and so are less susceptible to wooding however there is therefore a potential risk of saturating surrounding receivers. The drawback is that much of the energy transmitted is "wasted" as it radiates equally in all directions. To achieve high data bandwidths considerable energy is required so often these aerials are not suitable for very high bandwidth requirements.

Directional antennae provide much high potential data rates as they only transmit their energy in one direction which helps considerably. The challenge though is ensuring that the aerials are not blocked or wooded by parts of the superstructure or other topside equipment. Placement is dependent on the role of the aerial. For satellite communications visibility of the sky from directly above to all around and at low elevations is important.

Typically to overcome wooding two or more antenna are used to ensure that regardless of the orientation of the platform to the target that sufficient coverage is maintained.

AESA type aerials may allow for repurposing aerials for multiple uses across transmit frequencies. The directional nature of these arrays could allow a network of them around a ship to provide seamless coverage from sea level to directly up and 360 degrees around the platform.

6.2 THE IMPACT ON THE RADAR SIGNATURE

A common response to adding new aerials is to add more sponsons or brackets to a superstructure. Whilst this may improve coverage it is also likely to add to the Radar Cross Section (RCS) due to the addition of new micro-geometries to the platforms structure. The addition of micro-geometries is known to perceptively increase the RCS of a platform and so should be considered when changing the upper deck layout.

7. BELOW DECKS DESIGN CONSIDERATIONS

7.1 NETWORKS

Whilst it is not always possible to identify the physical or functional requirements of a future item of equipment, there is a high likelihood that it will be networked either within its own systems boundary or to other systems.

During system design networks are usually optimised with some spare capacity based on the demands set out by the customer and refined by the System Integrator. Sometimes unused networks are added and these could be copper cables (CAT 5 or similar) or fibre optics. Due to the small size of individual fibre cores dark fibres (i.e. initially unused ones) are left to provide additional redundancies and capacity for future growth.

A further consideration could be to consider the likely compartments that connections are required and ensure that network access ports that can be re-configured exist. Allowing the patching and re-routing of networks around the ship would also provide greater flexibility. Including node rooms or routing spaces to allow for additional switches or routers to be added to setup new networks or sub-nets would provide a low cost way of providing considerable options for future networked capability enhancements. The precise location of these spaces should also be carefully considered.

Providing easy access for patching and routing would enable the infrastructure installed at build to be re-configured. Changes in technology mean it is likely that network technology may become obsolete and need to be replaced. By considering this at build and providing access to maintenance spaces and network cables would allow for networks to be added or removed at a lower cost by avoiding running networks in inaccessible places like along the keel or outside the pressure hull.

7.2 SHARED COMPUTING INFRASTRUCTURE

The addition of high capacity storage arrays and large servers is enabling computer power to be shared between capabilities and systems. As technology develops this infrastructure will become quickly obsolete but this needs to be accepted and the design of platforms needs to embrace the need to change hardware easily.

The shared computing environments could handle many of the transversal requirements such as secure storage, environmental management more efficiently and more effectively than a federated architecture of dedicated computing but managing functional safety may increase costs and mission criticality of the system may be subject to a risk if single points of failure are introduced. Assuring and accepting complex and interrelated systems can increase the burden of trials with increasing needs for regression testing on all elements when a change is made.

The System Integrator needs to consider the system architecture to consider efficient trials and acceptance without it becoming burden.

Sharing consoles would also need to be considered as many functions can be displayed on any screen but some functions, particularly safety critical ones, need to be accessed from certain positions during physical installations like safe fire switches.

Establishing a robust security architecture particularly in relation to information and cyber security will be supported by separately infrastructure and handling security as an enabling function. Managing security collectively will reduce the risk of security breaches and reduce procurement costs by allowing projects to leverage security protection and enforcement measures taking place at a platform level.

8. CONCLUSION

The pace of technological change will continue to increase and with improvements and the ubiquity of networking in commercial applications and equipment makes the adoption of these systems more common on naval platforms.

Considering infrastructure on warships and submarines as an enabler rather than a necessary part of some systems will provide a route to maximise the capability of the "Fight" system and the support of these platforms.

On the upper deck providing greater consideration of communications infrastructure such as aerial locations, mutual interference alongside other sources such as sensors is needed during the early phases of platform design.

Below decks networking, making use of shared computing infrastructure and common consoles will enable the introduction of software only capabilities running on common hardware but also provide architectures to provide transversal functions such as cyber security on behalf of other systems.

9. REFERENCES

1.J. Gambetta and S. Sheldon, "Cramming More Power Into a Quantum Device", https://www.ibm.com/blogs/research/2019/03/power-quantum-device/, 4 March 2019.

2. I del Portillo, B Cameron and E Crawley, "A Technical Comparison of Three Low Earth Orbit Satellite Constellation Systems to Provide Global Broadband", 69th International Astronautical Congress 2018, 1 Oct 2018.

3. W O'Neil, "The Cooperative Engagement Capability (CEC); Transforming Naval Anti-air Warfare", Case Studies in National Security Transformation Number 11, 2007.

4. L Girffiths, J Rigby, "Future Air Defence: Countering Emerging Threats", RINA Warship 2018, 11 Sep 2018.
5. R O'Rourke, Navy Lasers, Railgun, and Gun-Launched Guided Projectile: Background and Issues for Congress", Congressional Research Service, 23 Oct 2018.

9. AUTHORS BIOGRAPHY

Simon Willmore holds the current position of Head of Mission Systems at BMT. He is responsible for oversight of the Mission Systems capabilities as well as leading on the platform design for naval warships and auxiliaries. His previous experience includes a career in the Royal Navy as a Weapon Engineering Officer and as part of BMT has supported the T26 Global Combat Systems setting up a GFE management, led as the combat system platform lead for several platform designs from auxiliary tankers to mine countermeasure vessels and frigates as well as providing consultancy support to the UK MoD and other governments internationally.

Chris Burnside holds the current position of Principal Mission Systems Engineer at BMT. He is currently leads the T31 combat system platform design team as part of Babcock's Team. He is responsible for managing the interface between the platform design and the combat system integrator. His previous experience includes roles as a system design authority on the QEC project and systems designer for submarine equipment at a major OEM.

Elizabeth Sutton holds the current position of Mission Systems Engineer at BMT. She is an experience CAD and design engineer who has supported numerous platform design projects as well as an engineering projects adding or removal capabilities from Royal Navy warships and submarines.

THE USE OF CAD SYSTEMS TO MANAGE MODULARITY IN MULTI-ROLE WARSHIPS

Rodrigo Pérez Fernández, SENER, Madrid/Spain, rodrigo.fernandez@sener.es
Erno Peter Cosma, SENER, Madrid/Spain, erno.peter@sener.es

SUMMARY

Modulation requires new approaches to the entire shipbuilding process, general arrangement for the warships need careful analysis of interaction and physical connection between modules: electrical, water, air conditioning, etc. and between modules and the elements surrounding them.
The analysis of areas with high potential for modulation will lead, by the application of the criteria decided, to a selection of several standard modules to be designed in *CAD* systems for those cases where it is expected to achieve significant savings by producing series of modules.
The appropriate application of the modularization theories demands a difficult exercise of coordination among the design of all the units, systems, zones or disciplines present in the compartments and also direct implications in constructive strategy which will require specific protocols for the correct implementation of the modules during the constructive stages.
The implementation of new tools to handle modules in warships is particularly important since it remarkably simplifies the analysis, what leads to a high potential in the improvement of the coordination among design areas and disciplines, and consequently to a more efficient design.
The typical outfitting and equipment modules for these areas shall be defined, interfaces to the equipment and specification of cables, pipes and air conditioning systems needed for the module. The goal will be to define a limited number of standard modules, which allows exploiting serial effects and containing a significant part of outfitting in order to achieve savings from pre-outfitting.
But the concept of modularization has even a wider meaning for warships when thinking in modularity, not only oriented to the design and constructive advantages but also in the future evolution of these kind of ships in order to be reoriented easily towards different uses, future upgrades or changes that should be realized to be able to undertake different kind of missions thanks to an efficient *Plug and Play* system.
The aim of this paper is to develop concepts and design solutions for modular machinery space, on-board automation and warship equipment. To achieve those possibilities, not only the modularity concept will be necessary, it also needs a clear vision of which will be the lifecycle of the warship and its present and future purposes or possible upgrades in order to create a design. *CAD* systems are essential tools to provide the vision and the base to design and simulate the different possibilities that a warship would be able to provide thanks to the application of a modular concept during its complete lifecycle.

NOMENCLATURE

AI	*Artificial Intelligence*
ASW	*Anti-Submarine Warfare*
CAD	*Computer Aided Design*
CFD	*Computational Fluid Dynamics*
GCS	*Type 26 Global Combat Ship*
IoT	*Internet of Things*
IP	*Interim Product*
IT	*Information Technology*
LCS	*Littoral Combat Ship*
LHD	*Landing Helicopter Dock*
LSS	*Logistic Support Ship*
MCM	*Mine Countermeasures*
MFTA	*Multifunction Towed-Array*
MOD	*Minister of Defence*
MRCS	*Multi-Purpose Combat Ship*
NATO	*North Atlantic Treaty Organization*
PPA	*Pattugliatore Polivalente d'Altura*
RFID	*Radio Frequency Identification*
RHIB	*Rigid Hull Inflatable Boat*
ROV	*Remote Operated Vehicle*
SUW	*Surface Warfare*
SVTT	*Surface-Vessel Torpedo Tubes*
TEU	*Twenty-foot Equivalent Unit*
UACV	*Unmanned Aerial Combat Vehicles*
UAV	*Unmanned Aerial Vehicle*
USV	*Unmanned Surface Vehicle*
UUV	*Unmanned Underwater Vehicle*
V/STOL	*Vertical/Short Take-Off and Landing*

1. INTRODUCTION

In the engineering field, the modularity concept, understood as an approach for the processes improvement of construction in diverse kinds of industry, has attracted the attention of naval architects, designers, manufacturers, ship-owners, managers and researchers.
The adaptation to the shipbuilding environment, set in the range of highly complex ship design and construction for high-intensity scenarios, drives the value of this study, which seeks to explore the different applications practices that has had for the optimization of productive employment in the construction of warships, as well as the decisive future having modularity in this field (Jacobi, 2003).
It is possible to state that it was during the 1960´s when the modularity concept began the adaptation as a transversal concept to different industrial fields, establishing it as a hierarchical system consisting of

independent subsystems. Since the 1970's the modularity has taken relevance in the maritime world. In this way, these subsystems complementary concepts are added such as standardization, adaptability and flexibility to display the modular scheme implemented in the industry.

The projection of the industry towards the implementation of the modularity as a basis can be understood to achieve improvements in the supply chain, optimizing considerably the time of construction and configuration, allowing its adaptation to the different missions and tasks to perform quickly and efficiently, becoming a growing trend that the navies of the world are implementing.

But while technology and optimum advancement of processes is of great importance, a global reality led to the optimisation of resources should be viewed. These should be done, addressing a complementary issue that has served as a support for the justification and use of modularity as a necessity, taking into account the scenario faced by different navies to declining budgets and production costs ever higher. It should be noted that, clearly different political and economic overall levels, the regional environment of each nation, leading in some cases to an industrial base with limitations of development within the environment of shipbuilding globally.

Obviously such vastly generalized economic behaviour leads to that each navy must strive increasingly to reduce the costs associated with the design of warships, as well as the production, acquisition, operation and modernization of their vessels.

2. EVOLUTION OF THE MODULARITY CONCEPT

Modularity has received during the last years a lot of attention from all industries trying to increase both cost efficiency and customer satisfaction. The term module was however originated in ancient times, when it was given an architectural or structural meaning. Later, along with the industrialisation the term was attached to system parts performing distinct technical functions. Nowadays modularity is most of all seen as a way of handling complexity. From the business point of view the concept has a strong inclination towards standardisation and economies of scale on one hand, and customisation and variability on the other. One of the most important criteria of modularity can thus be said to be the functionality of the module during different stages of its lifecycle (Perez et al., 2015).

Originally the term *module* had a purely structural meaning. Last century the term became ever more used for functional distinctions within a product or system. More recently the term has been used to perceive immaterial things such as knowledge, business processes and organisations. Below it is presented the basis for all of these ways to use the concept of module.

The term module is derived from the Latin word *modulus*, which was a unit of measure in classical architecture. The concept can be regarded to originate from the work of roman architect Marcus Vitruvius Pollio (Vitruve) who worked under the Emperor August.

The idea of combining standardisation with functional design received more attention only in the beginning of the 20th century through the paradigm of functionalism when the module was linked to the concept of a building block. It was soon realised that the productivity of building was greatly improved when the building components (modules) were prefabricated, instead of making them on the building site at difficult locations and object to unpredictable situations.

The functionality of the building block was not directly connected to the module, as the module still was only related to the geometry of the interface. More recently, the building block concept does not necessarily refer to modules as they can also refer to standard or parametric design such as arrangements, system schematics and/or set of engineering procedures. Miller and Elgård (1998) categorise the drivers behind modularity according to: creation of variety, utilisation of similarities and reduction of complexity; or in other words (mass) customisation, standardisation and clarification.

Apart from balancing standardisation and customisation many see modularity as a means for handling complexity (Langlois, 2002). Attention was devoted to the real interdependencies between modules. By minimising these interdependencies the work on each module could be done quite independently of each other; i.e. the information needed to carry out the works within one module were hidden from those working outside that module. First of all he distinguished between visible design rules and hidden design parameters and he continued by claiming that modularization is beneficial only if the partition is precise, unambiguous and complete.

Whereas the previous point dealt with the concept of modularity from both an engineering and a theoretical point of view, this point seeks to explore the practical economic realities attached to modularity.

As described, the idea of creating variety through modular production was presented already some 40 years ago (Starr, 1965) and well implemented in the car manufacturing industry. The central idea has been to meet specific customer requests at near mass production efficiency.

Other authors claim that mass customisation is so similar to many other contemporary concepts, such as lean production, agile manufacturing and concurrent engineering, that a clear boundary cannot be drawn between them. In practical terms, mass customisation can be seen as the production strategy or business concept evolving from a flexible, modular product and process design. A modular design enables the configuration of modules and processes into customer unique deliveries, which as a whole constitute the basis for successful mass customisation.

Mintzberg (1983) have developed a framework to describe the three types of standardisation used for coordination in organisations: standardization of work processes, standardization of output and standardization of skills and knowledge.

A central concept in mass customisation is configuration. Configurators are tools for achieving customer integration, e.g. the idea of utilising the customer as a co-designer, as

they enable fast and realistic visualisation and simulation of different product configurations.
Quite recently lifecycle engineering has become an important concept in management and engineering. Lifecycle engineering approaches seek to ensure optimal functionality throughout all the stages of a product and therefore becomes important for companies that pursue a modular design strategy, since the functionality requirements in different phases of the lifecycle of a product are likely to have implications for decision concerning the product architecture. Ishii (1999) lists four perspectives on functionality of modules:

- End user perspective.
- Engineering for Production methodologies.
- Maintenance perspective.
- Recyclability perspective.

And just finishing this point, Ulrich (1995) suggests that changes to the product are most easily accommodated through modular architectures, due to the drive for minimisation of interdependencies in such systems.

3. THE CHALLENGE

The concept of multi-role vessels seems to be the answer to the challenges of the future surface fleet (Majchrzak, 2017). It is possible to see examples of the importance of this concept in some of the most powerful Navies, and a growing number of designs that includes the capability to embark and employ modular mission payloads (Abbott, 2006).
The capability of some of those ships that could be already built or simply offered, depends almost entirely from the embarked payloads, while other vessels have a wide range of equipment that will be complemented and expanded by embarking payloads.
The *Black Swan* concept is a *95 meters* ship of *3,000 tons* with a core crew of eight sailors and space for *32 mission* specialist. It is described as a mothership vessel whose operational capabilities relay completely on the payloads that would reach *400 tons* and an equivalent of *20 containers* in volume. Those payloads will be installed on the mission deck and will be individually accessible and connectable to the ship services.
As an important part of the design, the ship will have a large hangar and a *Chinook-capable* flight deck.
It is supposed that one of the main advantages of this concept is the cost: four *Black Swan* plus mission payloads could be built for the cost of a traditional large escort vessel.

Figure 1. Second trimaran ship of the *LCS* Program docked at *Austal* shipyard, Mobile, Al (USA). Source: *U.S. Navy*

In the *United States Navy* (U.S. Navy) it is possible to find another example of ship that can be configured for different roles by changing mission packages (each mission packages provides the seaframes with capability for a single or *focused* mission), the *Littoral Combat Ship* (LCS) designed for operations near shore.

In early 20th century, small ships and transformable concepts to warfare began to be attractive and the Navy started to work on it by adopting features from the *Street Fighter* ship envisaged in the 1990s by Vice Adm. Art Cebrowski, head of the *Naval War College*. Key factors for the *street fighter* were to be a cheap, agile, fast and small (less than *1,000 tons*) vessel, capable to work in missions where big warships demonstrated bad behaviour. Inspired by this *street fighter concept* the *LCS* was converted into a reality but the final result was not what the original *Street Fighter* was meant to be, as a consequence extra requirements that had to be considered into the design.
Those modifications in design requirements while *LCS 1* was being built, lead to expensive reconfigurations and as a result, besides the extra cost, the final design of the *LCS*:

- Have almost non-existent growth margins.
- Mission packages must weight no more than *105 tons*.
- Accommodation for the crew was necessary enlarged because more crew was needed to accomplish the missions than those initially estimated.

These factors, combined with a poor autonomy and limited armament and sensors, make *LCS* an example of the limitations that this kind of modular concept can imply if the decisions taken in the design are not the correct ones. These wrong decisions are also the result of the complexity of making and managing important redesign during the construction which makes the project execution even more difficult.
Nevertheless, thanks to this first approach, the *U.S. Navy* identified that some of the modular payloads should be permanent components of the ship, so it is a critical decision to identify what should be part of the modular

concept and what should not. This decision will condition all the operative life of the vessel.
In 2012 *BAE Systems Naval Ships* won a contract to design *Type 26 Global Combat Ship* (GCS) for the *Royal Navy* that will replace the *Type 23* frigates. The *Type 26* is conceived as a highly capable and versatile multi-mission warship designed to support anti-submarine warfare, air defence and general purpose operations anywhere on the world´s oceans with its *149.9 meters* in length a maximum beam of *20.8 meters* and *6,900 tons* of displacement.

Figure 2. Mission bay being used to disembark a *RHIB*. Source: *Babcock* investor presentation

One of the most important parts of the concept of operation for this ship is the design of the mission bay, conceived for quick loading and unloading operations of equipment modules, vehicles, boats *UAV/USV*'s or stores. This arrangement allows modules or other cargo to be landed onto the flight deck and then moved through the hangar to the bay.
This important part of the design was initially designed to be housed underneath the flight deck, but as the design matured, it was moved to a more central position.
This bay, combined with the flexible accommodation provision, is a good answer to achieve other functionalities that will determine the usability as a multi-role vessel.
The crane system, based on a model used for handling containers on North Sea oil rigs is rated at *15 tonnes* and can extend to the side of the ship for loading and unloading operations, and it was necessary a lot of experimentation to determine how it can be effectively modified to accommodate a range of movement and orientation of the ship.
Without the crane, the mission bay is no more than an empty space and it will allow, for example, to launch and recover autonomous unmanned systems.
Type 26 has an important difference as regards *Payload* modularity from the *U.S. Navy LCS*, ships main sensors and weapons are fully integrated.
The *Type 26 GCS* project team is also leading on a couple of projects that will benefit *NATO* standardization, and one of them is the module interfaces.
In 2007, *BAE Systems* started to talk about *UXV* Combatant, a combination of warship and next generation carrier for unmanned craft concept.
The basic design of the *UXV* combatant was the *BAE's Type 45 Daring-class* destroyer that runs on diesel-powered electric turbines and includes two *164-foot* decks joined in a V shape, from where unmanned vehicles will be launched using electromagnetic catapults and ramps.
One of the advantages of these unmanned vehicles launched from *UXV*, that could determine the design, is a smaller space for the take-off because the acceleration could be higher than the acceleration needed for a manned vehicle limited by the human capacity to support such acceleration.
Take off and land management will be supported by technologies like *RFID* tags and radars.
UXV has some differential characteristics:

- Could be operated effectively with a minimal crew of *60* (enough for three shift changes) where in other similar ships it is talking about hundreds of sailors. Some additional crew members specialized to operate the drones will be added.
- Versatility is a key factor for the design. To achieve this, *BAE* engineers work in this concept with the American defence contractor *General Dynamics* on a separate project called "modular mission bays", a set of *Plug and Play* features that would enable commanders to quickly *re-role* the boat for different purposes:
    - Submarine hunter.
    - Minesweeper.
    - Platform for supplying ground troops.
    - A launching pad for drone-based air strikes.

This concept will allow for example, to store the drones in other ships or land bases and when the *UXV* is called for a mission that requires some of those drones, the drones needed for that specific mission will be dispatched to it.
Besides those *Plug and Play* kits, the *UXV* will include on the foredeck:

- Missile batteries that could house both surface to air and ship to ship missiles and cruise missiles.
- A large calibre gun that fires six-inch munitions.
- A *155 millimetre* medium-calibre gun.

Types of drones supported.

- *UAVs*: Unmanned Aerial Vehicles.
- *UACVs*: Unmanned Aerial Combat Vehicles.
- *V/STOL* aircraft: airplane able to take off and land vertically or on short runways.
- Helicopters.

Nowadays *UXV* is still a concept. Other interesting examples can be found in the *Italian Navy*:

- The *Pattugliatore Polivalente d'Altura* known by their acronym PPA, an innovative multifunctional ship that replaces the *Lupo/Soldati class frigates* and the Minerva class corvettes and can be considered as a Multipurpose Offshore Patrol Ship.
- A *Logistic Support Ship* (LSS).
- A multipurpose amphibious unit *Landing Helicopter Dock* (LHD).

The high innovation level of these three classes of ships is oriented towards high efficiency, flexibility and capability of re-configuration in order to be able to assume different mission profiles, particularly for standard military purposes and for civil protection and rescue at sea operations.

PPA is a highly flexible ship designed to serve multiple functions ranging from patrol with sea rescue capacity to Civil Protection operations and in the *Full* configuration it will be able to operate as first line fighting vessel.

Figure 3. Artist representation of the *PPA*. Source: Ufficio Stampa - *Marina Militare*

With a length of approximately *143 meters* and a full displacement between *4,000* and *5,000 tons* will be capable to operate high-speed vessels such as *Rigid Hull Inflatable Boat* (RHIB) thanks to its lateral cranes or a hauling ramp located at the far stern of the ship.
Different configurations of combat systems will be available:

- *Soft* version for the patrol task: equipped to self-defence.
- *Full* version: oriented to a complete defence ability.

At the stern and centre of the ship, two modular zones allows the embarking and handling of various types of containerized modules:

- Stern area:
  - Up to five modules in *Twenty-foot Equivalent Unit* (TEU) containers of capacity.
  - Two side openings are part of the design and will add flexibility to the procedure to deploy unmanned vehicles and will be also used as *Rescue Zones*.
- Central area:
  - Up to eight *TEU* containers and/or *RHIBs* up to *15 meters*.
  - A powerful crane to handle boats and containers.

This ship will be procured in multiple configurations:

- *Full*: All sensors and weapons will be installed.
- *Light+*: This configuration will be able accept the systems that includes the full version as required.
- *Light*: Will be prepared to accept systems included in full version, but not all of them.

The crew varies between *90* (for *Light* configuration) and *171* (for *Full* configuration).
An interesting innovation of this ship will be a *Cockpit* system developed by *Selex ES* and *Fincantieri* that uses augmented reality to allow an effective sailing management and combat system operations with fewer operators.
Another plan is to develop *Google Glass-type* glasses linked to external cameras, which would allow navigators to have a *360-degree* view of the outside of the ship.
The *PPA* can be considered more like a traditional frigate, putting a lot of focus on the ship's own sensors and weapons rather than on modular payloads, but it is possible to see how the *Italian Navy* consider the approach to this modular concept.
Finally, the *German Navy* and its *Mehrzweckkampfschiff 180* (MKS 180) *Multi-Purpose Combat Ship* (MRCS) programme proposal, describes a future multi-mission platform with built-in modules designed for specific military missions that will confer them multi-mission capabilities.

Figure 4. *MKS 180* graphic concept of the *Bundeswehr. Source: Bundeswehr* (*unified armed forces* of Germany)

The *MKS 180* missions:

- Self-defence and combat missions.
- Creation of a maritime picture above and underwater.
- Maritime surveillance and embargo control.
- Military evacuation in crisis situations.
- Escort for merchant ships.
- Command of a naval task force.

The basic version of the *MKS 180* is already a full-fledged combat ship and interchangeable components supplement this core capability and specializes the vessel for specific roles with mission modules like:

- *Anti-submarine warfare* (ASW) module that turns the *MKS 180* into a dedicated submarine hunter with on-board helicopters and their own sonars that works combined with the allied maritime patrol aircraft and submarines, securing a large area against submarine threats.
- And a custody module that turns the ship into a floating base for anti-piracy missions that includes an additional sanitary station that can be used under quarantine conditions.
- Module equipped with a diving chamber and other special equipment for mine warfare.

The mission modules will be accommodated into three different areas:

- A flex deck located below the flight deck at the stern.
- Two additional flex decks located halfway along the length of the superstructure.

One key requisite for the *German Navy* is that the replacement and commissioning of the modules should be executed as fast as possible and anywhere (shipyard not necessary), without interfering with the ship's structure.

Another key requisite for the modules is to withstand the climatic and oceanographic conditions that could be the tropics or the polar waters.

4. DRONES AND MODULAR CONCEPT

As can be seen in the examples above, the use of drones in the military world is an important part of the present and the future of the warships. Warships should be designed considering that they should be able to house and operate different kinds of drones that require:

- Operative space.
- Space for the complex systems that could be required for the control and operation of these kind of vehicles (winches for *Remote Operated Vehicle* (ROV) umbilical, take-off platform for unmanned aerial vehicles, etc.).
- Accommodation for the crew specialized to operate them.

It is possible to consider the payload composed by each drone and its control unit as a payload kit that also needs a specific infrastructure on-board with specific bays and facilities that have to be installed as a *Plug and Play* module.
The presence of unmanned motherships is also taking more and more relevance and some of the ships that nowadays might not be considered state of the art war-firing vessels can have a new role as unmanned motherships, at least this is an idea that is being commented.
A key factor is the quick evolution of the unmanned vehicle technologies and this is one of the reasons that moves to the modular concept, the capacity of replacing obsolete systems by new ones with the latest technologies in an easy and fast way.
This kind of payload kits are not so demanding as other kind of kits that requires almost a fusion with the baseline ship, but depending on the Drone and its contour requirements, the *Computer Aided Design* (CAD) systems can also be the tool to determine the best way to optimize the baseline ship with the necessary bays to accommodate the complete system and to determine the best strategy to allow a fast *Plug and Play*.
Drones are, maybe, the future that will finally decide about how the modularization should be defined.

5. PAYLOAD KITS: THE PLUG AND PLAY PAYLOAD

Navy Capt. John Ailes, program manager for the littoral combat ships´, mission modules commented on *Military.com* in April 2012: *"The modular nature of LCS means almost anything an engineer can dream up, within reason, can work. Meanwhile, shipbuilders Lockheed Martin and Austal USA don't need to pause construction because the Navy wants a new sonar, or other new equipment; instead, by moving in parallel, the Navy can phase things in as they become available".*
This concept *Payload kits* refers to those payloads that bring capabilities to those baseline ships to convert them into high capability single or multi mission platforms configured to face specific complex scenarios or missions. The payload kits are designed to plug into the existing services connections in the hull, which would supply chill water, data connections and electrical power buses, all of them will be in specific locations ready to be uncapped and fitted to the kit´s component modules.
In the *LCS* case, the ship can be configured for different roles by changing mission packages.
Those packages includes mission module equipment (weapon systems, sensors, etc.), carried craft and mission crews.

- *ASW*: detect, classify, locate and prosecute enemy submarines.
- *Mine Countermeasures* (MCM): Detection and neutralization of mission threats.
- *Surface Warfare* (SUW): maritime security and prosecution of small boat threats.

An example of payload kit could be the one designed for *ASW* configuration and consists of:

- *ASW* kit with a *Multifunction Towed-Array* (MFTA).
- *Sonar* and *SLQ-25 Nixie* towed torpedo decoy (already in their own self-contained modules) fitted in the payload bay aft under the flight deck and around the stern ramp.
- *Mk-32 Surface-Vessel Torpedo Tubes* (SVTT) are fitted to the weather decks.
- This kit could be supplemented by the undersea unmanned vehicles designed for use with the *LCS ASW Mission Package.*

Modules were designed to allow a single *LCS* to change roles in a matter of hours, (in 2014, *Independence* switched from mine to surface warfare modes in *96 hours* on short notice) at any commercial port allowing it to quickly optimize effectiveness against a threat.
To simplify shipping, storage and availability of correct handling equipment, the Mission systems fit inside standard *TEU* containers or on *ISO* compliant flat racks and transport containers.
The *Mission Packages* reconfiguration will occur at homeport or overseas, using the pre-positioned Mission Packages or those that have been transported by air, land or sea and staged near the operating area.
To understand what is a *Mission Package* here is a clear description:

- Mission systems: *Vehicles + Sensors + Weapons.*
- Mission Module: Mission systems + Support Equipment.
- Mission Package: Mission Modules + Mission Crew + Aircraft.

To achieve this flexibility, the Navy is developing and procuring mission packages to meet the joint warfighting requirements. The quantity of each mission package type differs based on an analysis of projected operational needs, therefore, mission packages are developed and procured separately from the LCS seaframes.

Another important component is the Mission Package Support Facilities (MPFS) that plays the integral role of primary provider for all seaframe-embarked mission packages, float maintenance issues, provide technical support physically and virtually via distance support service that works *24x7* to all deployed mission modules and will ensure LCS´s ability to reconfigure mission packages according to operational demand.

The modularity key factor, besides the possibility to configure the same ship for different missions, is to keep technologically updated along of *30 years* that a ship of this type could be in service and it is possible to do this easily by using standardized payload bays.

Systems whose technological evolution can be predicted and identified as something that could remain obsolete in a short time, like weapon systems, radar and Sonar, etc. are the best candidates to be converted in modules that can be replaced in the future by new versions with better technologies that will ensure the best behaviour in missions.

A periodic evaluation of the requirements from the combatant commanders in order to ensure the correct amount of capability available in the fleet at a given time is another key factor to determine the design and production of new payload kits or possible modifications. When the correct requirements are applied, logistics, manpower costs, training and maintenance costs will be optimized.

But, is there a limit for modularity?. As it is mentioned in the *MOD Joint Concept Note 1-12*, dated May 2012, *"Systems Not Platforms. Much as torpedoes, submarines and aircraft changed the face of maritime warfare in the last century; unmanned systems will do the same in the 21st Century. In the future, unmanned systems could help to provide a solution for keeping a balanced fleet by matching the required capability to future threats, available resources and ordered tasks. This future concept would concentrate investment in systems, rather than the ship, and a change in emphasis to one that does not see the ship as the fighting platform".*

It is important to avoid seeking modularity at all costs because it can conduce to bad results.

It seems that the baseline ship should include at least the most difficult systems to transfer onto off board solutions:

- Its own air defence missile battery, with the appropriate suite of sensors.
- A main gun, for naval gunfire support.
- Light guns for close range self-defence.
- Ideally would also carry land-attack missiles.
- At least a hull mounted sonar for obstacle and mine avoidance at a minimum.

Without these systems, the ship will have serious survivability and capability limitations as *LCS* have demonstrated.

In the publication of the *U.S. Department of the Navy, Designing the future U.S. Navy Surface Fleet of Effectiveness and producibility* in 1989, it was described an interesting idea, the *Technology clusters* which would seek synergistic linkages within programmatic structures:

- Carriers of large objects: Common hull that would replace aircraft carriers, amphibious ships, and fleet auxiliaries.
- The large carriers will be complemented *Scout fighters-surface combatants* with common hulls.

This study clarifies the value inherent in using a single hull form and the efficiencies that can be gained by moving toward more common systems and construction.

There are also efficiencies that can be gained by streamlining combat systems, as well as man/machine interfaces into a single common-core system.

With this in mind, what is the proposed solution? The proposal consists in the design of modular *Plug and Play* operation systems supported by two different-sized baseline hulls moderately armed and manned.

- *Leviathan*: Larger hull capable to carry large objects with reconfigurable holds and flight decks. This concept would be based on the modern amphibious-assault ship hull. The current gas-turbine technology would transition to electric drive and the missions modules would fit within the hangar, hull and superstructure. In essence this version will be Cerberus class plus a large flight deck with the added capacity and volume to berth and feed more crew and passengers.
- *Cerberus*: a *Scout fighter* role, smaller than leviathan but equally designed as multi-mission craft. With this concept, if the commanders need a specific capability, Cerberus can be tailored to meet mission requirements with a minimally invasive yard period, instead of a lengthy modernization. The baseline, itself will be operative without any additional modules or modifications (mission package) and could be used as simple presence, theatre antisubmarine warfare or integrated air and missile defence requirements throughout her life.

The kits proposed for *Cerberus* and *Leviathan* are not assembled into the platform at the point of construction because that will mean a limit for the long-term adaptability of a given hull, the proposed kits would be large and would need an intermediate level maintenance facility to install (similar to the *LCS* mission modules).

The design of the modular systems and the interfaces would be designed considering that the *Leviathan class* could carry any or all of the *Cerberus class* while the *Cerberus* will be limited by the volume and size of modules.

The objective is a true reconfiguration ability and true modularity that goes further than the idea of decoupling the platform and the payload.

The challenge is how this is going to be done, and an important part of the answer can be found in *CAD* systems.

6. ADVANTAGES AND DISADVANTAGES OF THE MODULARITY IN THE NAVAL SHIPBUILDING INDUSTRY

Broadly speaking, the benefits of modularity are the same as clarification, standardisation and customisation. To be more specific, in the following paragraphs it have been listed some generic or usual advantages and disadvantages of modularity in the naval shipbuilding sector.

Advantages:

- Keep technologically updated during the service life: allows the vessel warfighting capability to quickly adapt to evolving threats using improved technology. In the case of changes in operating conditions and advances in technology, only the module may need to be modernized (Johnston, 1994).
- Multi-crew concept: This concept could reduce the overall cost of the ship and will allow a smaller crew who continuously operate and maintain its core systems.
- The scalability of the fleet lends itself to cost and time savings for manpower and training resources (afloat and ashore) since manning and training are delivered at the particular requisite level.
- Example: If a combat system is not installed on a platform, there is no need to have personnel or supplies on board to do maintenance.
- Don't forget that people are the most costly expense for the military and one of the objectives that can be achieved with this modular concept is the crew reduction. As an example *DDG-1000*, *Zumwalt class*, call for a crew of fewer than *150*, whereas in its 1992 configuration, the *New Jersey* required a crew of almost *2,000* sailors. Except for *Nimitz-class* aircraft carriers (with *3,200*) no ship in the *U.S. Navy* approaches that many crew.
- Standardized payload bays to make an easy module replace/installation is a key factor. Those interfaces should be designed to be compatible with new and future modules.
- With a single hull design it is possible to meet the widest variety of missions.
- Just with the baseline concept, the ship will be able to assume missions.
- Greater stability and sea keeping, and more space around vital machinery, weapon and command components.
- If a system is not necessary for a specific mission, it will be better to be removed, and the modular concept allows it. It means less weight, which in turn provides additional fuel savings and streaming range.
- Unused mission modules can be stored and maintained independently of the ship.
- Using the modular payload kits, those kits can be quickly and cost-effectively upgraded in yard while the ship is still operative for other missions.
- A true modular design gives the fleet the ability to tailor the mix of baseline ships, single mission ships and multi mission ships to be exactly what the combatant commanders require.
- The modules do not have to be procured for each ship and can also be purchased independently, any time.
- The single hull increases numerous efficiencies and complicate adversary targeting them.

Disadvantages:

- The power generation supply: The necessary energy for the baseline warship operation will be substantially lower by itself, but it will be necessary to dimension the power plant of the ship for all possible present and future modular payloads combinations. The range between each situation will be much wider than in warships designed and optimized for specific purposes and this will mean an important challenge for engineers to design a power plant able to work with the best performance in each case.
- In designs like the *Black Swan* concept, space imposes less accommodation room for the crew and also for the teams of specialists that operates the unmanned vehicles. This will mean an important effort in accommodation design in order to accommodate them as best as possible.
- The *LCS* has problems derived from the immaturity of the modular payloads, when the development of the payloads started after the development of the ship and in a period where unmanned vehicles where still primitive.

Other considerations:

- Challenging concept (both technically and financially).
- The cost of the *Black Swan* concept because of the extreme reliance on external systems, mainly unmanned vehicles like *UAVs*, *USVs* and *UUVs* needed will be moved from the ship to the unmanned vehicles and the rest of modular systems.
- A key factor is to be able to change the modules as quickly as possible.
- Periodic evaluation of the requirements is necessary for the modules design.

7. HOW SHIPBUILDING CAD TOOLS COULD HELP?

It is easy to understand all the advantages of the modularity and the creation of a common frame (baseline) ship to bed the different payload kits depending on the mission, but a wrongly oriented design will lead to a ship with lower skills than a dedicated one, created and optimized for a specific purpose. This is why the design tools and the engineering work will even have a more critical role in these kind of projects.

First of all, the design of a ship based on this modular concept requires a very strong study of all the possibilities/missions that the ship should face, both for the baseline and for the modules; which means a big challenge for engineering teams that will define the design, but it will be possible and reachable not only by applying knowledge and experience but also with the help of the right tools.

The design of the baseline ship will be directly conditioned by the payload and depending on each configuration, the baseline ship should be designed to work in the best conditions talking about stability, sea

keeping, power plant performance, etc. Fit all the possible conditions under the same baseline hull, especially if it is not able to determine how the payload kits will be, seems to be a non-easy resolution problem, and this is one of the challenges that this kind of ships should face.
It means that the design of the baseline ship is one of the critical aspects of the concept because this will basically be the same during the entire life cycle of this kind of vessels. For this reason, *CAD* systems are essential tools to analyse and determine the best baseline ship for the different present and future possibilities of configurations for the different missions.
Besides this fact, *CAD* systems will allow to determine how the modules and the connections should be, in order to fit to the requirements of the different payload kits that have to be controlled and fed with energy, water, oil, data, etc. and a clear critical point for all this job is to define standardized connection docks to connect the modules with the ship.
CAD systems will also be essential to plan how the modules should be plugged and unplugged by using the capability to simulate the different scenarios and foreseeing the possible problems.
As mentioned before, one of the key factors for this kind of vessels is how fast can be reconfigured. To achieve this, it will be necessary to define a living guide or manual for a fast *Plug and Play* process and a training program for the shipyard operators in order to follow the steps and reach the desired configuration in the fastest and most secure way.
It is important to remember that this *Plug and Play* processes will be not necessarily developed in a shipyard, with examples like *MKS 180 Multi-Purpose Combat Ship* where one of the design requisites is that the replacement and commissioning of the modules could be executed anywhere.
It has been defined as *living* the guide or manual because during the operational life of the ship, it is expected that new modules will be designed and produced, and each module can include substantial differences with the others and it could mean modifications in the guide. The well-known concept of *Digital Twin*, extensively used nowadays, could have an important relationship with this live guide.
The training systems should include the interaction in *Virtual*, *Augmented* and *Mixed Reality* environments for a faster and more intuitive learning process introducing concepts like gamification which has being demonstrated as an excellent way to increase the effectiveness of the learning processes.
Beyond all this, mixed reality will have a really important role from all points of view, especially for maintenance operations in general and operations with modules in particular. As example it is possible to look at how the *Italian Navy* has important plans for this technology using it to control the ship from the cockpit.
Mixed Reality will be the human interface with a system that links the real asset with the geometric model generated with a *CAD* system, both synchronized following the *Digital Twin* concept which will be also synchronized with its current status by using *Internet of Things* technologies (IoT).
If analytic tools that explode the historical data are added in combination with *Artificial Intelligence* (AI) oriented to provide a diagnosis of the health of the system, or there is access to that previously mentioned *live* manual of *Plug and Play* protocols for the modules, there is going to be a powerful instrument that nowadays can become real thanks to the enormous evolution of all the technologies involved in these processes, all of them supported by the key technology, the communications infrastructure, each day more and more powerful.
As can be seen, *CAD* systems are in the middle of all this ecosystem that will be connected to multirole vessels during the design process and their complete operative life, being the application of these tools decisive for a successful design and a successful operative life.
Another important problem where *CAD* tools can help is in the important decision to define a system as an integrated part of the baseline ship (conventional solution) or transform it into a modular system. This is a crucial decision because, depending on what it is chosen, it could mean a problem for the baseline ship if the decision is not correct; maybe this is the most important reason why the design of this kind of concept is by far more difficult than the design of a conventional ship.
Once the decision is taken and it is decided to modularize something, the design of the connection dock is also a decision that determines the future of the ship and here takes an important role the definition of that docking points as standards.
The design of the spaces and the mission bays for the modules are an important factor too, not only the spaces to plug them and operate on-board but also the necessary spaces for all the operations that should be performed to load or download, including the movement on-board to reach its final position.
Modules spaces and mission bay should be designed carefully and foreseeing all the future attachments because once defined, the future modules will be constrained by the adopted design of those elements.
It has seen how for *BAE's Type 26* one of the most important parts of the ship is the mission bay that includes a crane system allowing fast load and unload process. The design of that kind of crane is a challenge where *CAD* systems are necessary, and with simulation makes a perfect combination to achieve the best design.
The design of the crane could be also used to create a simulator for training purposes in order to optimize the load/unload process.
Fit the mission systems inside *TEU* containers like it is possible to see in the *LCS* case is a good option when it is possible, but in case of dimension all the spaces and mission bay for this dimensions without an extra margin, it should be clear that for the future modules this will be the limit.
In any case, if future modules have more size, the best solution seems to be to separate the module in pieces and then assemble them on-board, but this will mean an extra time to prepare the ship for the mission.

8. TOOLS TO OPTIMIZE MODULAR CONCEPT

It is clear that there will be some key questions that will determine the modular concept to be applied in a new multi-role warship.

During the project stages, main questions that should be answered could be:

- What will the different missions to be covered with this ship be?
- How should the baseline ship that will be the different present and future payload kits be?
- What should be modular and what should be integrated?
- Which will be the design and characteristics of that modular present and future kits?
- How should the docks that will connect the modules with the baseline ship be?
- How should be the definition of the strategy and logistics to enable an "as fast as possible" swapping process to prepare the ship for the different missions by changing the payload kits?
- How should be the on-board infrastructures and equipment designed to make possible a quick loading and unloading operations of equipment modules, vehicles, boats *UAV/USV*'s or stores, like the bay designed for the *Type 26 frigates* of *BAE Systems Naval Ships*?

During the operational life, considering the contour conditions of the baseline ship, the project is still alive and will experiment a continuous evolution thanks to its modular condition:

- New modules will be designed to qualify the ship to face different missions or to modernize the initially designed modules.
- Continuous improvements in logistics, yard and off-yard operations and protocols to optimize *Plug and Play* processes.
- Minimize and optimize operations to keep up to date the baseline ship.
- Training and maintenance of the modules and base-line ship.

In the centre of the key tools that will help to optimize the design of a multi-role warship will be the *3D* model, and that means that the *CAD* systems are the protagonists. All these tools and technologies, combined with experience, deep knowledge, and a visionary point of view in order to foresee the technologies that will be implemented during the operational life of the ship, will be the perfect mixture to obtain the best results, the perfect multi-role warship capable to face with success different type of missions.

Nowadays, there are relatively new concepts and technologies that will complete the circle, or old ones that now have come back with strength thanks to technology improvements, and in centre of the circle, as it was mentioned before, is the model:

- *Digital Twin.*
- *IoT.*
- *Virtual*, *Augmented* and *Mixed Reality*.
- *AI.*
- *Big Data.*
- *Data mining.*
- *Data analytics.*
- *Machine Learning.*
- *Simulation.*

Digital Twin is a widespread concept nowadays, and it is clearly one of the best allies for the multi-role warships.

Digital Twin means the connection between the real asset and the design, the virtual world, and it seems to be a natural connection, but it was not possible until some other technologies evolved enough.

IoT, *AI*, *Machine Learning*, are now the technologies that make possible a real effective *Digital Twin*, especially the *IoT* technology which could be considered the basic connection between *Reality* and *Virtual World.*

There will be a continuous and bidirectional data exchange between real asset and the model in order to keep updated both entities and to reflect the reality in the virtual world and inverse.

The flow in the direction from *Real Asset* to *Virtual* model consists in a combination of sensors and persons sending data. The sensor usually sends the information thanks to its connection to the communication infrastructure while the human operators communicates with the virtual model with apps normally designed for devices that allow mobility and wireless connection for the workers into the infrastructure.

The inverse flow usually comes from a control centre that manages the data provided from the real world reflected into the virtual design and has two ways to communicate with the real asset: through interaction with widgets created in the digital model that will send the direct order to a fiscal actuator or by transmitting an order to a human operator.

Both communications can be done by a human or an automatic process if the virtual asset includes *AI* or an automatic program that simply has an output depending on the input parameters.

Machine Learning could be also an important component of the *Digital Twin* by learning from the data, the previously adopted decisions and the consequences of those decisions. When the system has enough criterion to adopt its own decisions, it will be able to suggest possible solutions or send alerts thanks to the capability to make predictions.

The final goal for a *Digital Twin* is to reach a near real-time status and to achieve this it is necessary to learn and updates itself from other sources, in addition to the previously mentioned: from other similar machines, similar fleets of machines and from larger systems and environment in which it may be a part of.

But what does it happen with all the information that flows? The digital twin also integrates historical data from past machine usage to factor into its digital model and this information can be used to create predictive models, detect patterns, early detection of problems, etc.

In an interview done in May 2019 to James Geurts, the *U.S. Navy*'s acquisition chief, explains how the *Digital Twin* is the answer for an easier upgrading of the ship's combat

systems. On the other hand, cybersecurity will be a key factor because *Digital Twin* concept depends directly on network. James Geurts also commented that the *Digital Twin* is also applied over shipyards.
There are industries like automotive or aeronautical industry where the *Digital Twin* is a widely applied concept, nevertheless, shipbuilding industry is still starting with this concept. Under our point of view the sector where *Digital Twin* has more sense is the Military, in particular, multi-role warship is the most interesting application because the *Digital Twin* fits perfectly with the multi-role concept due its continuous evolution during the complete operational life.
Once assumed that *Digital Twin* is the perfect concept for this kind of warship, it is important to understand how this *Digital Twin* should work for the Industry but it will be mentioned it in brief later on.
Initially, only the model will exist, a design made with *CAD* system.
Recall the importance of the baseline ship and the decisions to be taken to design it considering the different possible missions.
With *CAD* systems it will be possible to design this baseline ship, but maybe it will be not possible to decide a direct way to the objective and it will be necessary to create different designs that run in parallel while there are not enough criterion to decide.
Imagine a situation where we start to create the design of the structure of baseline ship and it is not clear what should be the dimensions and the position of an operation bay and we should decide between different options. If a situation like this happens, *CAD* system should be able the create a branch whose origin is the structure that's already designed and from this tree trunk we should be able to develop the different options for the operation bay.
The possibility to create branches should be unlimited and it should be possible the interconnection between the components of the design in order to avoid unnecessary duplicity into the design process, for example: if a branch is created with two possible solutions for the operation bay, it will condition the design of the affected part of the structure but the rest of that structure does not necessarily be different. The same could happen with a crane designed for the operation bay, the bay can be in a different position but the crane component could be the same, and if it is necessary to make a change in the crane, this change should not be applied twice (one for each branch) but just once and the change will be effective for both.
The final result of this process will be a single project but with a number of variants depending on the path followed through the branches.
This is similar to the applicability concept developed by FORAN® system, but with applicability in FORAN®, the result will be a number of different and complete projects with shared and synchronized parts.
Applicability was born to create different versions of a ship, and the concept described above generates a single project with variants, and this fits perfectly with the modular concept that is about a baseline ship (the tree trunk) and the payload kits (one or more payload kits can generate a branch).

Once this project with variants is mature enough, we can study the advantages and disadvantages of each option in order to decide the best one.
Obviously, this way to design the ship requires more work than the necessary to design a dedicated warship, but is worth it if it is considered that with this work it will be able to decide about critical aspects in much better conditions and minimize the possibilities of constructing the wrong baseline warship.
During this process, simulations will be done, as well as tests with models that will be executed on the different variants of the ship, to analyse structure (finite elements), hydrodynamics (*CFD's*, tests in the model basin), stability, power plant dimensioning, etc., nothing different from other ships, but all these activities should be done considering the different conditions of the ship, depending on the modules used for each configuration.
In any case, this won't different from other type of vessel designed to work in different load conditions. Maybe the only difference could be the margins to be able to assume payload kits not already designed, whose weight, dimensions and position will be determined in the future, when the ship has been already built, but also depending on the standardized connection docks.
There are other types of simulations that should be done, which are particularly important for this modular concept (for each ship variant):

- Simulations to analyse and define the *Plug and Play* process and operation of the different modules.
- Simulations to analyse the evacuation process.

The first type of simulations should include:

- Load/unload process: operations inshore/offshore with the intervention of cranes, other ships, helicopters, etc. that could carry or bring the modules to be installed or uninstalled in the ship.
- Movement of the modules between the origin and the connection dock.
- Movement of the module already connected if module has own movement.
- Sailors' movements and activities to be performed during this operations.

There are tools in the market to simulate industrial processes like assembly lines, but there is still room for improvements by making a closer relationship between the *3D* model (even under development) and the simulation, and maybe the key factor is the world of game development.
Each day is clearer that platforms for game design have a huge potential for engineering purposes. For example, real time *3D* development platforms, designed initially for multi-platform game creation, are now extensively used for architecture purposes.
There are proposals that offer a platform focused on Architecture, Engineering and Construction, with which it is possible to create a connection between pure engineering, the *3D* model created with a *CAD* system, and all the power that game development platforms that offers: physics, materials, lights, animation, and much more.

With this base and the potential to create applications for different devices: *PC*, *Smartphones*, *tablets*, *Virtual*, *Augmented* and *Mixed Reality* devices that will work with our *3D* Model, a world of possibilities will be opened.
Once this connection is established, the *3D* model can come alive into a virtual environment and develop applications to generate simulations in a virtual world by assigning to the *3D* models movement characteristics and interaction into a realistic environment.
Imagine that all the possible variants of the ship can be tested into a virtual world to analyse the possible problems of each design. From this analysis there should be able to generate incidences for the detected issues with tools designed for that purpose and send them to the *CAD* system to modify the design or just reject it.
All these steps will form a dynamic cycle of optimisation that will provide a better understanding of each design and a way to reach the final baseline ship and then, off course, the modules, with a process that minimizes the risks of a failed design.

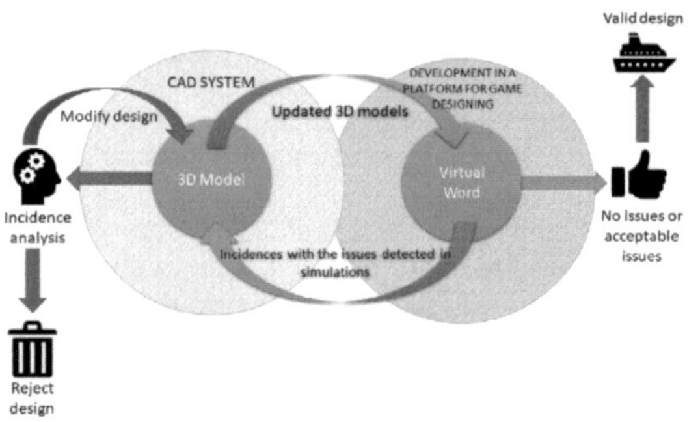

Figure 5. Diagram that explains how could be the workflow between the *3D* Model of a *CAD* system and the applications designed with a game development platform

Evacuation is a high priority part in the project, and for this kind of warships there is a particular characteristic that there have seen in the previous examples: the different number of sailors that will be on-board depending on the configuration of the ship. This variation in the number of sailors could affect the evacuation plan and should be designed depending on the configuration of the warship for a specific mission.
There will be sailors that should be prepared for the evacuation processes for all the different possible configurations, because they are the crew for the baseline ship and, on the other hand, there will be specialists whose knowledge about the evacuation process will be reduced because his presence in the ship will be just for some specific configurations.
The co-author of this article has developed a concept that is particularly interesting for this kind of ships in his Doctoral Thesis (Peter, 2014). There are several technical papers (Peter et al., 2014a and b) that talk about this model and the *R&D&i e-flow project* developed for land infrastructures that was created by using the model detailed in the Thesis (the project was also under the management of the author of the Thesis).

His work treats about an innovative system designed to manage the evacuation process like an event under constant evolution during the incident that causes the evacuation.

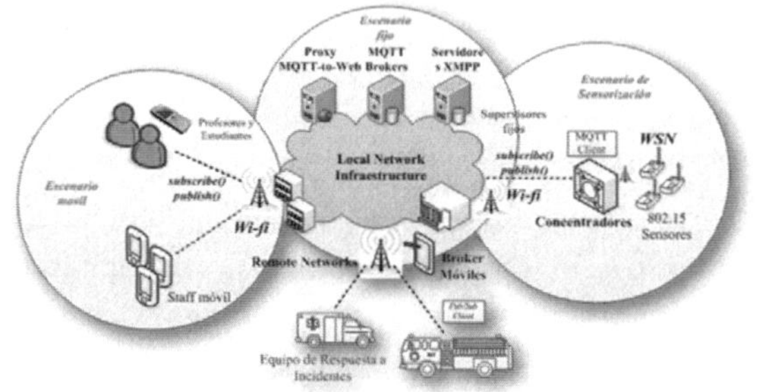

Figure 6. Schema of the *e-Flow* project. Communication system for user notification (Morales et al., 2013)

Nowadays, the evacuation plans are predefined, according to rules and without considering the variability of the conditions in real time. To make it possible, the Thesis involves tools, new algorithms, devices and infrastructures that manages dynamically signals, analyse the situation of the infrastructure dynamically by acquiring the information provided from a dedicated infrastructure that provides information from sensors.
The core of this Thesis is to develop a *3D* evacuation simulator created directly in a *CAD* system by scripting, with two group of tools:

- *e-Flow Net Design*: is used to create on the *3D* model the elements that will be used for the simulation, areas (rooms, halls, etc.) and access points between areas.
- *e-Flow Simulator*: this group of tools is used to define obstacles, signals, etc. and finally to define the characteristics of the simulations to be carried out.

This model includes the concept of dynamic signals and a human model behaviour based on the information provided by the scenario through the signals, geometry, and human models in the scene.
Two key factors in this model are the field of vision of the human and the uncertainty about the route, which make it different from most of the evacuation simulators, which only determine exactly the route that should be followed by the human model.
With these kind of tools for evacuation simulation, it is possible to design an evacuation plan for each possible configuration and this is also an important part of the process as it could mean changes in design.
Another part developed in this thesis is the use of virtual environments for training in evacuation processes and how it could be possible to learn in a virtual environment what should be done during a simulation process.
If the idea of create a connection between the *3D CAD* Model and the game development platform is applied, it will be possible to create a virtual environment to simulate the warship with all the different configurations and make the crew skilled into this virtual world to:

- Get familiar with the vessel.

- Create virtual alert conditions with fire, smoke, etc. and training how to behave under those situations.
- Implement the concept of dynamic signalling system,
- Interact with other crew members into the same virtual ship.
- Interact with crew controlled by Artificial Intelligence.

All these possibilities could be implemented and create serious *Virtual Reality* games that should include tools for analytics, evaluation, scenario configuration, etc.
The best thing is that all the effort done to create these big world of possibilities does not finish when the warship is built, all this job will evolve with the *Digital Twin* concept:

- To the design of new modules.
- To a continuous improvement in the processes of changing the configurations for new missions.
- To an optimized training system for evacuation.
- To the design of optimized evacuation plan, or to get familiarized with the ship in its different configurations.

All this systems are based on the *3D* model that, particularly for this kind of ships, it gets more importance than any other, because the multi-purpose vessel principle is the variability on the design and this will be possible with a *3D* model that evolves with the ship in a continuous update process that involves more technologies besides those previously commented like the laser scanner that will have an important role for the *Digital Twin* concept.

9. THE ADAPTABILITY CONCEPT

It is necessary to put into context the concept of adaptability and their relationship turns with the main theme of this study therefore to make clearer the position where modularity is in basic naval principle, before exploring different environments that leading logical efforts to the understanding of the issues in engineering, therefore must describe the capacity of adaptation of a design projected into a future of technological development and systematic change of needs, which arise at the same time two concepts such as modularity and flexibility (Pahl and Beitz, 1996). In the Figure 8 it is possible to visualize the relationship established by *RAND Corporation* for the adaptability concepts and which must be taken into account for better understanding (Schank et al., 2016).

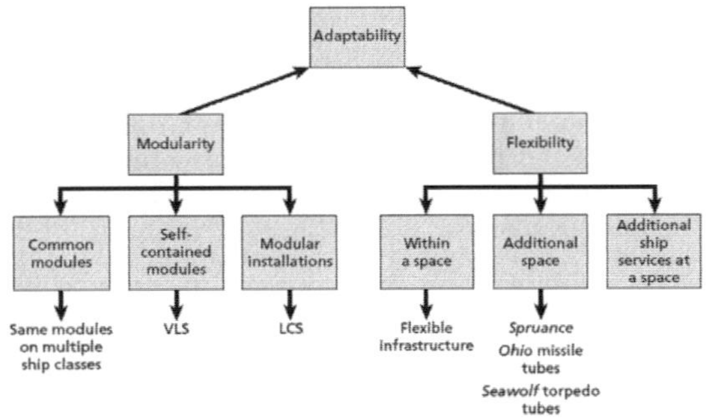

Figure 7. Adaptability and its relationship with the modularity and flexibility (NATO, 2004)

The term modularity is generally used in three different ways. In the complex design of engineering with naval application systems, the term is used in connection with interchangeable units to the same space station modules. Thus, in regard to the construction and the modularity of the architecture refers to the development of systems by standardized components. In manufacturing, the modularity has to do with use of interchangeable units to create variants of products (Ericsson, 1999) (Largiader, 2001).
In the surroundings of naval architecture is said to a complex product is *modular*, therefore it is divided into several components and then these are assigned to modules in architecture or specific plan. Modularity in an engineering perspective is usually used for three main purposes (Baldwin and Clark, 2004):

- To make manageable complexity.
- To enable the work in parallel.
- To adapt to future uncertainties.

The former purposes are used in the conceptual level of modularity and apply to the features required in the naval shipbuilding industry.

10. CONCLUSIONS

The modularity in the naval shipbuilding industry is a growing trend, which has been implemented successfully since the early stages of design.
This paper has focused to specify the concept of modularity, understood transversely to the flexibility. This is possible when deriving in turn from the adaptability, subject to which it is decisive to establish performance requirements, restrictions on the design and the margins of growth right from the initial phase of every ship project.
Each of the different types of modularity provides differentiating attributes shipbuilding and design, both in the operational performance of each platform.
This is how the first type, associated with the concept of shipyard construction modules, optimizes the time and resources to build, since it allows to develop different modules in parallel to reach a final integration of the entire structure of the vessel; the second type associated with operational performance through the concept of modules of mission, although it must be interpreted and applied from in-phase design, incorporating an additional factor

that enables the possibility of a dynamic configuration of vessels, to be able to meet the needs of an army with various areas of operation and their consequential tasks. However, that must be versatile, optimizing limited resources you have at the same time and determining a fast reconfiguration that makes this concept a differentiating factor in the operational performance of the vessels.

These two concepts are a highly complementary trend, since with them, optimizing the time and resource in the construction, and in turn it has configurable units that meet the modern defence challenges. However, it is relevant to conclude the parameters and common attributes that make it possible to implement modularity. As the case of design, should be viewed in the flow of constructive modules or constructive strategy and its tasks in parallel and predecessors to optimize construction time and reduce costs at this stage. Also, in standardization in the components to be used both in constructive modules as in mission modules, since they must meet the requirements of connection, connectivity, power and cooling required for its implementation, either in the time integration as configuration changes and then the maintenance during the life cycle of ships. It again addresses the issue of margins of growth that must be declared from the conceptual phase and future phases, which should be available for the configuration space, as well as which should be seen for future adaptations in three important edges which would be incorporation, modernization or replacement of systems, components, or new or already existing components without affecting the characteristics of the vessels.

Modular concept on a common frame is applied to multiple sectors: Car companies, Aeronautics,... but there is a risk of sub-optimization that can be minimized with a good design, but it is necessary to assume that the risk always exists and it must understand that a common frame for multiple purposes will never reach the optimization of a dedicated ship, but it is necessary to leverage pros and cons and decide; *CAD* tools can also help to adopt that decision.

If the objective of modular concept application is to adapt quickly the combatant commander´s requirements and the design does not allow a fast switch process, the advantages of the modularity will be not reached.

To achieve this fast switch process not only the payload kits and the docks are key factors, but also the general disposition of the baseline ship and the accessibility that should be designed to facilitate the swap process should be studied, optimized and defined in detailed protocols that will guide the operators during the process.

Without good logistics, modular concept cannot be applied if a fast adaptation is required.

Shipyard workers and the warship crew should be especially trained for these scenarios.

Maybe the decisions to be adopted will be less risky if the design consists in a baseline ship optimized for an specific purpose and be complemented with bays for payload kits that could enhance the ship to be able to face other kind of missions in addition to the one that specifically designed for.

This concept partially loses some of the advantages of the pure modular concept, but instead, the fear of having a sub-optimized ship will disappear and now it is possible to talk about an optimized ship with extra capabilities configurable through payload kits. Is this concept a better solution? In any case it will be a more expensive solution. It is not possible to have a single answer for all the cases, but depending on what it is necessary for the solution will be among these three possibilities:

- Dedicated ship.
- Modular concept.
- Dedicated ship with enhancement possibilities thanks to payload kits.

11. REFERENCES

[1]. Abbott, J. W. (2006): "Modular Payload Ships, 1975-2005". ASNE Engineering The Total Ship (ETS) Symposium, May.

[2]. Baldwin, C. Y. and Clark, K. B. (2004): "Modularity in the Design of Complex Engineering Systems".

[3]. Ericsson, A. and Erixon, G. (1999): "Controlling Design Variants-Modular Product Platforms". Society of Manufacturing Engineers, Dearborn, Michigan.

[4]. Ishii, K. (1999): "Modularity: A Key Concept in Product Life-cycle Engineering". Handbook of Life Cycle Engineering Concepts, Models and Technologies, Kluwer Academic Publishers, Boston.

[5]. Jacobi, F. (2003): "Modular ship design". Naval Forces 1/2003.

[6]. Johnston, R. (1994): "Operation: From Factory to Service Management". International Journal of Service Industry Management, Vol. 5, No. 1.

[7]. Langlois, R. (2002): "Modularity in Technology and Organization". Journal of Economic Behavior & Organization, Vol. 49 (1).

[8]. Largiader, C. A. (2001): "Modular Platform Based Surface Ship Design". Cambridge MA: MIT Libraries Document services.

[9]. Majchrzak, L. B. (2017): "Real Options Valuation in the design of future surface combatants". Naval Postgraduate School. Monterrey, California: Dudley Knox Library.

[10]. Mintzberg, H. (1983): "Structure in fives-Designing effective organizations". Prentice Hall, Engelwood Cliffs.

[11]. Morales, A. el al. (2013): "E-Flow: A communication system for user notification in dynamic evacuation scenarios". Ubiquitous Computing and Ambient Intelligence. Context-Awareness and Context-Driven Interaction. Carrillo, Costa Rica. pp. 207-214.

[12]. NATO Naval Group 6. (2004): "NATO/PfP working paper on small ship design".

[13]. Pahl, G. and Beitz, W. (1996): "Engineering Design–a Systematic Approach". Springer-Verlag, Berlin.

[14]. Perez, R. et al. (2015): "Modularity in shipdesign and construction". WARSHIP 2015: The Affordable Warship. Bath, UK. pp. 17-30. 10-11 June.

[15]. Peter, E. (2014): "Modelo Híbrido para la Ayuda al Capitán". PhD Thesis. E.T.S.I. Navales (UPM).
[16]. Peter, E. et al. (2014): "e-Flow - Sistema Integral Inteligente de soporte a la evacuación (1 part)". Ingenieria Naval. No. 924. pp. 87-95.
[17]. Peter, E. et al. (2014): "e-Flow - Sistema Integral Inteligente de soporte a la evacuación (2 part)". Ingenieria Naval No. 925; pp. 72-87.
[18]. Schank, J. F., Savitz, S., Munson, K., Perkinson, B., McGee, J. and Sollinger, J. M. (2016): "Design Adaptable Ships". Santa Monica, California: RAND Corporation.
[19]. Ulrich, K. (1995): "The role of product architecture in the manufacturing firm". Research Policy. 24(3).
[20]. U.S. Department of the Navy. (1989): "Designing the Future U.S. Naval Surface Fleet for Effectiveness and Producibility". Ship Production Symposium. Paper No. 24. NSRP 0310. September.

12. AUTHORS' BIOGRAPHIES

Rodrigo Perez holds a PhD in Marine Engineering. He is Head of the Naval Shipbuilding Area at SENER. He is also professor in several subjects at the Marine Engineering School of the Technical University of Madrid.

Erno Peter MSc and PhD in Marine Engineering. Works at SENER as part of the outfitting development team and coordinating mobile device tasks as part of the Industry 4.0 project. He is also a concept designer for unmanned vehicles and winner of several innovation and business ideas awards. His research fields also include evacuation processes, algorithms, simulation, etc.

NORWAY'S NEW MULTI-ROLE LOGISTICS AND SUPPORT VESSEL (LSV)

A J Aitken and **S R Jones**, BMT, UK

SUMMARY

This paper describes the key capabilities and features of the Royal Norwegian Navy's new multi-role Logistics and Support Vessel, HNoMS Maud, and discusses the benefits and drawbacks of a multi-role vessel considering some typical operational scenarios. It examines some of the design challenges and constraints encountered due to the multi-role nature of the vessel and the implications of standards on the design.

NOMENCLATURE

DSME	Daewoo Shipbuilding & Marine Engineering Ltd
FFBNW	Fitted For But Not With
HADR	Humanitarian Assistance & Disaster Relief
HIFR	Helicopter in-flight refuelling
HNoMS	His Norwegian Majesty's Ship
IPMS	Integrated Platform Management System
LO-LO	Lift-On / Lift-Off
LSV	Logistics and Support Vessel
NATO	North Atlantic Treaty Organization
NDMA	Norwegian Defence Materiel Agency
NorTG	Royal Norwegian Navy Task Group
OEM	Original Equipment Manufacturer
R&R	Rest & Recuperation
RAS	Replenishment At Sea
RNoN	Royal Norwegian Navy
SAR	Search & Rescue
VERTREP	Vertical Replenishment

1. INTRODUCTION

HNoMS Maud will soon enter service providing the Royal Norwegian Navy (RNoN) with a true multi-role Logistics and Support Vessel (LSV). She will be the largest ship to have flown the Norwegian navy's flag and will bring exciting new capabilities to the Norwegian Armed Forces enabling their contribution to a wide variety of national and international operations and giving Norway's armed forces a true blue water capability.

Maud will fulfil multiple roles, some concurrently and some alternatively. Her primary function is Replenishment At Sea (RAS) of solids and liquids but she also has a significant sealift capacity. She can be a mothership to the ships, submarines and smaller vessels of the Norwegian Task Group with facilities to support vessels moored alongside while offering the capability to support low-intensity maritime security/counter-piracy operations; Humanitarian Assistance and Disaster Relief (HADR) tasks; and peacetime support missions. Maud has a large NATO Role 2+ medical complex for up to 48 patients and a large aviation complex focused on NH-90 helicopter operations. Therefore, the LSV can fulfil the roles of a number of vessels found in other fleets, including AOR, forward repair, hospital ship and sealift. This mix of capabilities offers significant flexibility and value for money.

The ship was designed by BMT in the UK and is based upon BMT's Aegir® family of double hulled, twin skeg fleet replenishment ships [1]. The Norwegian LSV design carries the designation, 'Aegir 18NR'. She was built by Daewoo Shipbuilding & Marine Engineering Co. Ltd (DSME) in South Korea and, following delivery to the RNoN, she is due to enter service soon.

2. STRATEGIC CONTEXT & OPERATIONAL SCENARIOS

Norway has a 2650 kilometre long coastline bordering some of the world's most abundant areas for fish and energy (oil and gas). Its special relationship with the sea stems from 80% of Norwegians living within ten kilometres of the coast. The ocean has been the gateway to the outside world for more than a thousand years.

The RNoN is tailored for littoral and coastal maritime operations; meeting national as well as NATO and international demands for a rapid maritime response force (the country borders Russia and its main naval bases). Over the last two decades, Norwegian politicians have given priority to new military capacities and capabilities that will be crucial to Norway's ability to protect her sovereignty – on land, in the air and at sea. During the Cold War, all logistic support to the Norwegian Navy was either shore based from several naval bases or NATO alliance based. The transformation of the Norwegian Navy had to ensure a new fleet with a reinforced operational focus providing a versatile tool, ready to support Norwegian and allied security policy. Standardisation, Interoperability and Deployability were the prerequisites.

The LSV's main task is to contribute to Norwegian Navy's endurance in the new mission spectrum protecting the nation's Area of Interest acting as a logistics command platform, play a key role in NATO / UN task groups through providing military support, conducting boarding operations, and undertaking HADR operations. The requirements for the acquisition of a new logistics support vessel were set in order to:

- Improve endurance, flexibility and mobility of Norwegian maritime forces;
- Provide logistics support to maritime forces during operations and training, nationally and internationally;
- Contribute to deployment and sea basing capacity for maritime forces;
- Secure endurance of the maritime forces of minimum 30 days;
- Based on the vessel's inherent capacities, support the Army, Air Force, Coast Guard and civilian authorities with transport and medical support.

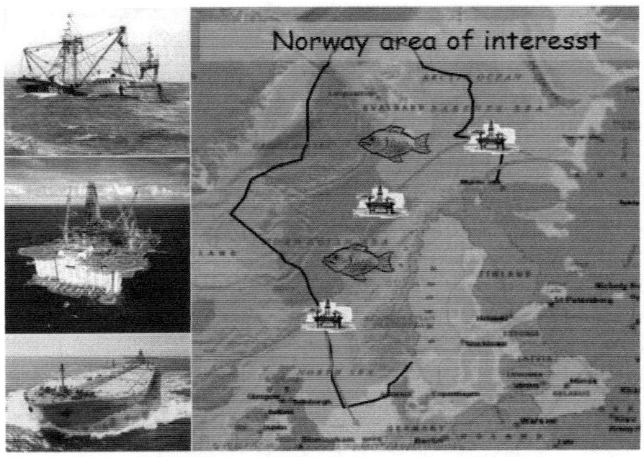

Figure 1: RNoN Area of Interest

The requirements were developed using the lessons learned from the Navy's support UNIFIL II off the coast of Lebanon in spring 2007, 'Operation Enduring Freedom' (Horn of Africa) and the humanitarian operations in Banda Aceh, Indonesia. [2]

HNoMS Maud will be capable of world-wide unrestricted operations and able to remain on station for up to 30 days without replenishment. One of her primary domestic functions is serving as a logistics carrier freighting military stores, vehicles and equipment between Norway's military bases dispersed along the length of its coastline. She must also serve as a "one stop" logistics support at sea away from Norway for the Fridtjof Nansen class frigates.

The above need resulted in the LSV requiring the capability to deliver the following roles:

Primary Roles:

- Replenishment at sea (RAS) of liquids and solids for NorTG and NATO vessels via abeam rigs, stern reels and VERTREP;
- Helicopter operations including landing and take-off of a wide range of large NATO helicopters plus level 2 maintenance of the ship's organic NH90 aircraft;
- Hospital Ship able to provide NATO Role 2+ care for up to 48 patients.

Secondary Roles:

- Sovereignty and authority through provision of an independent presence and interception of illegal fishing, including boarding operations using the ship's boats and/or helicopter;
- Sealift and logistics support to other units of the Norwegian armed forces;
- Support to civilian society using the ship's medical arrangements;
- Salvage and Repair;
- Search and Rescue;
- Humanitarian Aid and Disaster Relief;
- Pollution Control / Oil-spill preparedness;
- Command & Control and participate in network based defence.

Figure 2: LSV RAS with Nansen Class Frigate / Mothership with Ula Class Submarine

3. SUMMARY OF THE LSV SOLUTION AND HER ARRANGEMENT

The Aegir 18NR was developed to meet the Norwegian Defence Materiel Agency's (NDMA) requirements and is based on BMT's Aegir naval replenishment ship family. She has a large superstructure aft with liquid and solid cargo stowed forward in the hull. Above the cargo tanks and three deck magazine cargo hold is a deck of cargo solid stores and the cargo pipe trunk. This deck also accommodates the enclosed RAS machinery spaces. A double hull protects the fuel tanks and magazines with the magazines located as far from the accommodation as practicable. The aft superstructure provides improved visibility during RAS evolutions of the receiving vessel(s) and RAS deck operations when compared to AORs with forward superstructures.

The aft superstructure allows a large open deck to embark the deck cargo of up to 40 TEUs or other cargo which is serviced by the ship's cranes. The cargo deck is protected from green seas by the large bow visor, reported by the crew to have stopped all green seas from impacting the cargo deck during a crossing of the Pacific Ocean in heavy weather with 8-10m wave heights. Immediately forward of the superstructure are the combined abeam RAS stations capable of delivering F-76, F-44, fresh water and 2 tonne solid loads by heavy jackstay to port and starboard.

The aft of the ship is designed around accommodating the ship's large flight deck (able to land CH-53 or Chinook aircraft) and its hangar, which can house two NH-90 helicopters. The superstructure block contains the ship's command and control spaces, the hospital, galley and dining halls and the majority of the accommodation spaces. Immediately beneath the superstructure on 2 Deck is a mixture of the ship's stores, further accommodation and service spaces. Below 2 Deck in this region are the cargo pump room, the twin fully redundant main machinery spaces, auxiliary machinery spaces and the ship's tanks.

Figure 3: HNoMS Maud in Bergen

The LSV's principal particulars are summarised below.

Principal Particulars		**Cargo Capacities**	
Length overall	183.0 m	Liquid Cargo:	7,152 tonnes F-75/76 Dieso
Length bp	167.7 m		341 tonnes F-44 AVCAT
Beam	25.9m		273 tonnes fresh water
Depth	16.0m		49 tonnes lub oil
Full Load Displacement	27,500 t (end of life)		15 tonnes gear oil
Design Draft	8.6m	Solid Cargo:	400 tonnes containers or vehicles / boats
Design Speed	18 knots		200 tonnes munitions
Cruise Speed	~14 knots		30 tonnes cargo frozen / refrigerated provisions
Cruise Range	Minimum 10,000 nm		25 tonnes barrelled oils
Endurance	At least 30 days		4 tonnes distilled water (barrelled)
Crew	43 regular	Cargo Waste Reception:	77 tonnes sludge
Non-crew	116 embarked forces		37 tonnes bilge
Patients	48		34 tonnes black water
Total Embarked Personnel	207		117 tonnes grey water
			216 tonnes dirty water
			160m^2 solid waste

Table 1: LSV Principal Particulars and Capacities

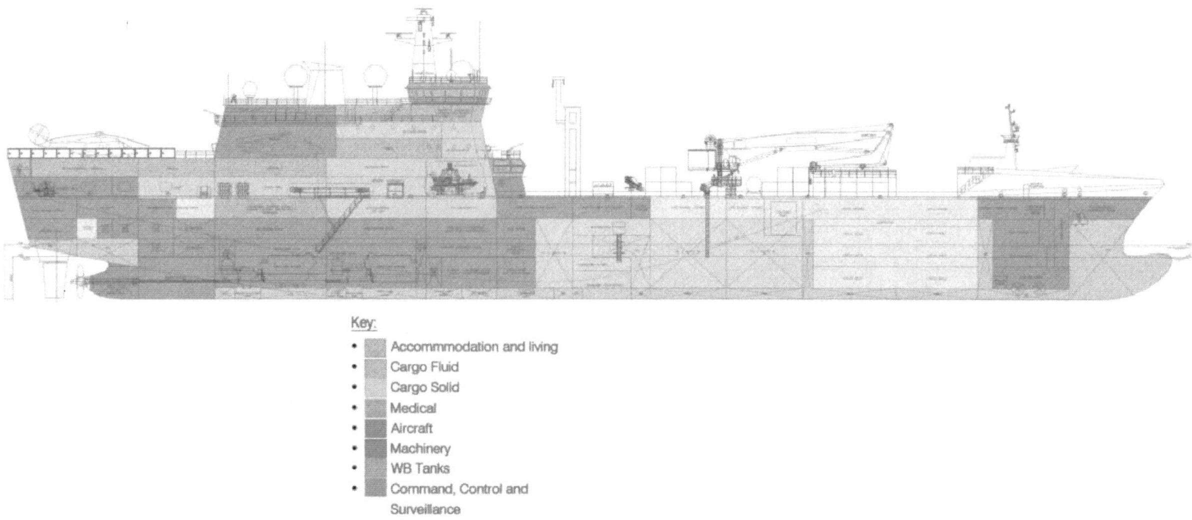

Figure 4: LSV Design Architecture

4. KEY DESIGN FEATURES OF THE MULTI-ROLE LSV

This section will discuss some of the ship's key features, which enable it to fulfil its multiple roles.

4.1 LIQUID CARGOES

The liquid cargoes are embarked in tanks permanently designated to their cargoes of Dieso, AVCAT or fresh water.

The abeam RAS stations enable the transfer of F-76 Dieso, AVCAT and FW simultaneously with vessels on each side of the ship. Astern 6" and 2 ½" reels allow underway replenishment in higher sea states and for replenishing smaller vessels.

4.2 SOLID CARGOES

The solid cargo stowage arrangements are intended to be as flexible as possible with internal and on-deck stowage provided. The magazine holds meet Def Stan 00-101 standards [3] and provide large spaces to embark various types of munitions as well as non-munition cargoes when required. The stores on 2 Deck comprise a mix of refrigerated/frozen stores, dry bulk stores and stowage of barrelled cargo. The magazines and some of the 2 Deck stores are outfitted with flow-forge and bin poles to provide flexible stowage and securing arrangements for different sized cargoes.

The ship's upper deck is designed to embark a variety of cargoes comprising of 10', 20' and 40' TEU containers, vehicles, boats and other stores. The deck can accommodate up to 60 TEUs with safe at-sea access provided to at least 50% of them. A matrix of twist locks and tie-down points are provided on the deck to maximise the quantity of cargo that can be embarked in the numerous loading permutations.

4.3 FLIGHT DECK

The single spot flight deck can accommodate a range of aircraft including NH-90, EH-101 Merlin, CH-47 Chinook, or CH53 Sea Stallion and is serviced with HIFR and on-deck fuelling facilities enabling inter-operation with allied forces. The hangar provides space for stowage of two NH90 aircraft or level 2 maintenance of a single NH90. A MANTIS aircraft handler was selected as it was considered to offer the most flexible solution – enabling in hangar manoeuvring to servicing and stowage positions.

Additional deck cargo could be accommodated on the flight deck but the space lost would limit aviation capability.

4.4 NorTG MOTHERSHIP

The LSV will act as mothership to the vessels of the Norwegian Task Group, particularly its smaller assets, such as the Skjold class corvettes, Atla class MCMVs and CB90 combat support boats. The ship is intended to provide alongside support to these vessels so is furnished with fendering and low level mooring fittings. Ship side doors (which also serve as the pilot embarkation points) allow access to the low freeboard vessels and the adjacent lobbies contain connections for services including electrical power, fuel, potable water and various liquid waste reception. The berthing positions are serviced by the ship's main and auxiliary cranes for cranes for the provision of stores and spare parts and reception of solid waste.

The support extends to enabling the boat and submarine crews to embark the LSV for Rest and Recuperation (R&R) using its laundry, gym and sauna (which were sized to service these additional personnel). The ship's galley will pre-prepare food for the vessels' crew to consume during their deployments.

4.5 HOSPITAL

The hospital is one of Maud's key capabilities. The permanent elements are able to deliver Role 1 medical care for at least 8 patients and can be expanded through the use of re-role spaces to provide a NATO Role 2+ capability. Its focus is to provide trauma care and is sized to handle a combat incident to a frigate-sized platform. However, the Utøya Massacre of 2011 was at the forefront of Norwegian minds during the development of the LSV's requirements so the RNoN envisages it will respond to a peacetime coastal or island incident if required.

When fully manned, the Role 2+ medical facility has a capacity for up to 48 patients and includes:

- 15 patient capacity triage facility close to flight deck;
- Resuscitation / trauma room for 3 patients;
- 2 table operating theatre;
- 8 bed Recovery, Intensive Care and protective Isolation (RICI) room;
- CT scanner;
- High and low dependency ward (non-crew lounge re-roled to high dependency ward, and officers' lounge re-roled to low dependency ward);
- Doctor's consulting room;
- Pharmacy & pathology laboratory;
- Recompression chamber for treatment of 2 diving casualties;
- Morgue with 10 places;
- Numerous supporting compartments / services including O_2 generation system.

4.6 CARGO & STORES HANDLING

The LSV's different roles require equipment and stores to be moved on and off and around the ship rapidly so design of effective stores handling systems and routes was key to the platform's effectiveness.

4.6(a) Cranes

The ship's Lift-On Lift-Off (LO-LO) capability is serviced by its 25 tonne SWL heave compensated crane; the position, reach and capacity of which allows it to service vessels or the quayside on either side of the ship. The crane is certified for ammunition handling and man riding.

Two small support cranes are also installed at the ship side for hose handling, storing duties and for servicing vessels moored alongside. They negate the need to operate the main crane to lift small items.

4.6(b) Side Ramp

The LSV has a sophisticated side ramp to allow loading and unloading of cargo by fork lift truck whilst in port. The ramp also allows ambulances to be brought into the ship to load/unload patients, which is essential given the ship's expected operating climate.

Developing the solution was a considerable challenge to achieve the large tidal range and variety of quayside heights specified. Side ramps are not normally fitted to ships of this type and accommodating the ship's draft range complicated the solution further.

4.6(c) Elevators

Several elevators facilitate movement of cargoes, stores and people between decks. A large munitions rated elevator (sized to be capable of lifting submarine torpedoes and the handling equipment) serves the forward holds and the main deck.

In the superstructure, a large personnel elevator enables the rapid and easy movement of trauma casualties with their associated medical teams to the hospital from the flight deck (02 Deck), boat positions (1 Deck) and 2 Deck transfer route. A further ammunition rated stores elevator links the flight deck to 2 Deck to facilitate the movement of VERTREP stores and munitions along the appropriate designated routes.

4.6(d) RAS Rig Design

In consultation with the RAS system supplier, the rigs' design was customised to provide free-standing independent systems with minimal footprint; thus maximising available deck space for unhindered cargo handling/storage and the ship's other roles. The solution also improves crew safety as moving RAS wires remain close to the mast minimising access restrictions and snatch-back zones.

The rig modifications also included designing for winterisation. Almost all the RAS machinery is below decks providing protection from the elements, and freeing up further upper deck space for cargo operations.

4.6(e) Logistics Routes

The main deck serves as external stores route for munitions, hazardous and oversized loads linking the forward and aft stores elevators (and flight deck to the holds). A "racetrack" arrangement is provided in way of the RAS rigs to enable unobstructed one-way flow. The cargo routes are forklift/sideloader capable.

An internal stores route on 2 Deck provides a weather protected route for transfer of personnel and non-hazardous goods. The route is sized for transfer of hospital patients on trolleys and palletrons carrying EURO-pallet sized loads.

5. DESIGNING FLEXIBILITY INTO THE SHIP

A single, multi-role platform is inevitably a compromise compared with multiple assets but a small navy does not have the manpower or may not have the budget to deploy numerous specialised ships concurrently. Also, despite the operator having a clear understanding of its need for the LSV and how it intends to use it; Maud offers the Norwegian forces several new capabilities and a significant step-up in capability in other areas so some roles will develop/evolve while the ship is in service. Significant effort was put into the LSV's design to maximise flexibility to enable the ship to fulfil its different defined and potential roles effectively while minimising the impact on capability.

This section explores how, through careful design and a very healthy collaborative approach, the arrangement and systems solutions adopted deliver the required flexibility and ensure the ship maintains the necessary readiness for re-role and, importantly, can be easily and rapidly re-roled.

5.1 FLEXIBLE POWER GENERATION & PROPULSION

The ship features the Aegir concept's Combined Diesel Electric or Diesel (CODLOD) hybrid power generations and propulsion system similar to that installed on the successful RFA Tide Class Tankers. The system comprises a twin shaft, fully redundant arrangement, with each shaft driven by a 7,500kW main engine, or hybrid machine in electric motor mode, coupled to a reduction gearbox to drive a controllable pitch propeller (CPP). Two further 3,300 kW generator sets provide electrical power. The hybrid machines can act as a shaft generators deriving electrical power from the shafts. This configuration provides a high degree of flexibility in operation; matching the wide range of modes and environments within which the ship must operate. The system optimises the use of the engines, improving fuel economy and reducing through life costs. The benefits of hybrid drives are discussed further in Reference [4].

A military benefit of the flexible power generation and propulsion architecture is the redundancy the system offers. The CPPs are coupled with high lift rudders and twin bow thrusters to create a very manoeuvrable platform. Its manoeuvrability and redundancy are essential to support the LSV's key roles as a replenishment or mothership and when operating alone to service remote ports without tug assistance.

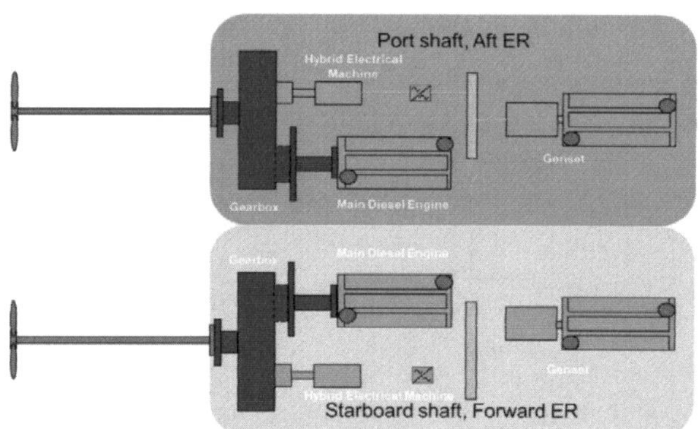

Figure 5: LSV's CODLOD Propulsion Architecture

5.2 SCALABLE MANNING

The LSV will be operated by different crew sizes depending upon its role. It can be sailed by a core crew of 43 (which can be reduced further when spending periods alongside). The ship has the capacity to embark an additional 116 persons for its most onerous missions plus 48 patients when the hospital is full, taking the maximum complement to 207 persons.

The RNoN's small size drives it to operate a very lean manned approach. As is common with small vessel crews, many of the personnel are trained to fulfil numerous roles and this rationale will continue to be used by the RNoN in the larger LSV; a concept the RNoN has proven with its frigates' scheme of complement.

To accommodate the range of complement, the compartment types are distributed as efficiently as possible. The interaction between each space and its function in each role was considered carefully to maximise the operability and habitability of the ship and minimise the need for multi-hatted staff to work in different parts of the ship.

The ship's systems are designed to be flexible and able to operate effectively across the range of operational scenarios including different levels of embarked personnel. Two boilers were fitted to provide tank heating, potable hot water and HVAC heating. These provide a level of redundancy and allow one to be shut down when the demand is low, rather than running a large capacity boiler at all times. Sections of the ship's HVAC system can be shut down, or operated at reduced capacity when those sections of the ship are unused. The galley is arranged so regions can be closed down when not required and the laundries were fitted with multiple COTS domestic appliances rather than large commercial machines. The laundry solution also reduces through-life cost: commercial machines require specialist support and/or service contracts whereas domestic appliances are readily available in most ports allowing adoption of an Upkeep by Exchange (UXE) policy.

To support the lean manned approach and the different numbers of personnel available, the RNoN is more ready to rely upon automation than some larger navies. Hence, the ship contains high levels of automation to allow it to function effectively across its different roles and many of the ship's systems and equipments are connected to the IPMS for remote monitoring and operation. The anchor windlass was selected to be operable from the Bridge – the personnel may not be available to stand up a full anchor party and it removes the need for an anchor party to remain on the fo'c'sle while navigating the confined waters of the high latitude fjords where the anchor needs to be immediately deployable. The flight deck nets are also power driven

5.3 MARGINS

Margins are incorporated into the ship's arrangement and its systems to ensure it functions effectively in each role. E.g. the mothership role requires significant margin in the accommodation for the NorTG vessels' crews: The laundries, gym, sauna and associated communal washing areas are sized for ship's staff and visiting vessels' crews. The Casualty Triage Area is a spacious compartment in the superstructure and can be outfitted with seating to serve as a recreation space for NorTG crews or during HADR missions, for example. The operations room was deliberately oversized to allow for future upgrades or temporary installation of mission specific equipment.

Weight and system margins are provided to allow for through life growth and modifications to the ship to be accommodated throughout her life.

Other margins were absorbed in the parent design because not all peak systems capacities are required simultaneously as a number of the roles are non-concurrent: High electrical load is needed during RAS but not at the same time as the ship is providing power to a frigate or submarine; capacity to receive and treat black water from NorTG vessels is required but the ship is provided with a system capable of supporting the full hospital.

6. MODULAR CAPABILITIES ON THE LSV

Some of the ship's capabilities are delivered through the deployment of modular solutions rather than being permanently available. Modular systems have been incorporated into ships with mixed results and experience has shown that if modular capabilities are difficult to embark/activate or are not present then the capabilities are not used. Alternatively, to ensure their use they may remain embarked permanently (e.g. the German Navy's modular hospital facility embarked upon the Berlin Class AORs). The modularised functions must be readily available or easy and risk free to embark and deploy. In the LSV's case, care was taken to only modularise the capabilities which were required occasionally; without immediate need; and where a low risk solution would result.

BMT considers modular solutions in two forms: "in a box" or more complex "Out of the Box" capabilities where individual equipment need to be installed and connected. The LSV employs both types.

6.1 "IN A BOX" CAPABILITIES

The ship's upper deck has a flexible lashing pattern to enable different configurations of cargoes, boats and containers to be embarked. The RNoN expects some of these containers to be modularised capabilities rather than cargo. The navy already operates a torpedo workshop from a 40' TEU and suitable locations are provided for it (it needs an 8m clear access in front of its doors). However, the NDMA foresees the LSV's roles developing through its life as the users realise the opportunities the ship offers. Therefore, five of the positions are served with connections to the ship's electrical power, compressed air, potable water, data and communications systems to facilitate embarkation of 'mission containers' that could be used for special operations, specialist stores, or even emergency accommodation. This also supports NATO's ambitions for modular capabilities [5].

6.2 "OUT OF THE BOX" MODULAR CAPABILITIES

The ship's hospital is an example of one of its "Out of The Box" modular capabilities. Modular systems are often touted as the answer to providing a medical capability on military vessels since they are needed infrequently and occupy a lot of space. BMT's conduct of the Joint Casualty Treatment Ship Capability Assessment Studies for the UK MOD taught us the benefits and short comings of modular hospitals, such as: the limited space available in TEU envelopes; achieving ship's fire protection and escape requirements; maintaining equipment in calibration and the modules' condition when not embarked; the logistics challenges and risks associated with shipping and installation/removal. BMT brought this knowledge to the design of the LSV's medical complex. The resulting solution is a predominantly fixed, integral capability to enable hygiene and safety standards to be met with expandable capability depending upon the mission, the embarked medical complement and the number of patients expected. The core trauma, surgical and intensive care elements are fixed, as is the pharmacy, laboratory, CSSD and doctor's consulting room. However, the high dependency and low dependency wards are delivered through re-roling the augmentees' and officers' lounges and a number of cabins.

Additionally, specialist facilities such as the CT scanner are provided as FFBNW capabilities. Dedicated, integral space is provided but the equipment, and specialist staff, are only embarked when the mission necessitates. This approach also supports ready upgrade of these large equipment items which have a relatively high technology refreshment rate (a lesson learnt through BMT's extensive work on the design of the UK's Primary Casualty Reception Facility (PCRF) on RFA Argus).

The comprehensive medical complex is concentrated over a single deck with a generous triage area for casualty reception located at flight deck level forward of the hangar. When the hospital is inactive, the triage space can be put to other uses, including, for example, a recreation space for crews of supported NorTG vessels. The triage facility requires little equipment to enable it to function so is easily re-roled. Conversely, the high dependency and general wards require a lot of equipment. Hence, the NDMA specified the re-roleable solution must accommodate all the medical furniture and infrastructure and be capable of full conversion within 24 hours. The solution places the beds and medical furniture behind a removable partition in the augmentees' lounge and the bedside medical systems connections behind removable panels in the bulkheads. The lounge furniture, carpets and the bulkhead partition are readily portable and will be moved into the cargo stores or empty TEUs. Both lounges are served by wheelchair accessible toilets and washrooms. The 6 re-roleable officers' cabins fitted for low dependency care are sized for trolley access and provided with oversized en-suite washrooms for easy access.

7. DESIGN APPROACH

The NDMA recognised that it had no previous experience of some of the capabilities so was reliant upon the designer and Original Equipment Manufacturers (OEMs) to help it define the solution. Consequently, an innovative approach was adopted whereby it was agreed to develop the solution jointly following contract award and positively encouraged engagement between the designers, operators, system suppliers and shipbuilder to develop the best solution available. The three key tenets were:

- Detailed Design Working Groups (DDWGs) – A series of workshops held throughout the design phase comprising the above parties (and often including the regulatory authorities) to develop an agreed solution through round-table discussions and addressing key design and operability features and safety obligations;

- The cost constraints and trade space within the contract were understood by the NDMA, shipyard and designer to allow "no cost" changes to be agreed.
- The NDMA was prepared to consider contract uplifts to achieve the capability where the DDWGs showed it was appropriate/necessary.

An advantage of the collaborative DDWG approach was that the participation of the end operators ensured the agreed solutions aligned with RNoN practice.

8. DESIGN CHALLENGES

8.1 ARRANGEMENT CHALLENGES

Despite the many benefits a multi-role vessel brings, there are inevitably some compromises compared with specialist, dedicated single role platforms. The key design challenge for a multi-role vessel relates to the architecture of the platform that is fitting all of the competing requirements into an arrangement that works well enough for all operational scenarios with priority being afforded to the most important tasks, or those requirements that have a safety or regulatory related need. The General Arrangement development needs to consider the usage of each compartment in each of the roles and manage the trade-offs between compartment adjacency and location within the vessel. This process is challenging and requires operational input from the end user, guided by experienced designers to determine the best layout. Inevitably, it leads to some compromise in a multi-role vessel to a greater extent than in a single-role vessel, noting that any ship design is a process of compromise to some degree. For LSV, some examples of compromise would include: the hospital complex would ideally be located further forward in the vessel to reduce motions, and would be located on the same deck as the flight deck and boats for improved access and evacuation.

A notable compromise on the LSV, is the lack of an enclosed clearway for the pre-staging of solid cargoes before a RAS operation. The need for an open and flexible cargo deck on LSV over-ruled the decision to have an enclosed clearway deck. The cargo can however be pre-staged on the cargo deck to increase the speed of RAS delivery. BMT has conducted studies into installing a clearway onto this design for another potential customer who did not have the need for such a large quantity of deck cargo, and has demonstrated this to be feasible.

The re-roleable lounges are relatively austere so their furniture can be removed quickly when required. When the hospital is fully operational and two of the three lounges are lost; the loss of recreational space will be felt by the crew. Personnel will need to spend R&R time in their cabins or the dining hall. However, Norway still has conscription in its armed forces so recruitment and retention is a less important consideration, enabling a lower accommodation standard to be provided for these personnel. Also, Norway's egalitarian society and the navy's flatter rank structure means combining officers and ratings facilities is not considered the compromise it might on other nations' platforms.

8.2 SYSTEMS DESIGN CHALLENGES

A major challenge resulting purely from the ship's multi-role nature was the design and specification of the HVAC system serving the re-roleable medical compartments. Lounges and hospital wards have very different heating/cooling and air change rate requirements which when coupled with the considerably different occupancy levels in each role made the demand upon the system very different. The design was complicated further by the need for the system to remain balanced under CBRN close down and open ship conditions.

8.3 SATISFYING THE REGULATORY FRAMEWORK

The Norwegian LSV is certified as a military vessel and is constructed to DNV GL's Naval Ship Rules, the NATO Naval Ship Code (ANEP-77) and the Norwegian military's NRAR standards as well as statutory standards including MARPOL. The nation does not operate civilian manned support ships so does not face the same challenge as the UK's RFA flotilla where the ships are registered as commercial ships fulfilling a military role so must comply jointly with military and commercial rulesets. Naval Flag administrations and relevant military stakeholders will accept tailoring of the naval rules and adoption of a risk-based approach more readily than civilian authorities. However, the tailored aspects must not compromise the minimum safety level of the vessel. The ship's diverse range of roles made it difficult to comply with all elements of the applicable rulesets so tailoring of the conflicting elements was necessary. Through BMT's involvement with the UK MOD's Naval Authority Group and the Naval Ships Assurance Service (NSASS) and involvement in numerous recent naval ship designs; its engineers have developed deep experience of tailoring naval rules to meet navies' bespoke needs so supported the NDMA with tailoring and gaining acceptance of the necessary elements of the rules applicable to the LSV.

HNoMS Maud is not certified as a tanker due to its relatively low volume of cargo fuels, which allows greater flexibility in the ship's arrangement and the installation of equipment making incorporation of its multi-role function easier. Nonetheless, to ensure that the ship does not present an environmental threat or safety hazard, the tanker features are compliant with MARPOL regulations and the key safety objectives of tanker rules. She is, therefore, a double hull tanker, physically, and meets meet all regulations for unrestricted global operations at sea and in harbour under IMO regulations and MARPOL requirements.

The ship complies with DNV GL's RPS (Redundant Propulsion and Steering) notation and the Maintain Capability requirements of ANEP-77 providing it with significant survivability. The key features include: military standard watertight subdivision and stability; extensive fixed fire-fighting systems; salvage systems; system redundancy and separation; and redundant and separate power and propulsion systems.

9. ACHIEVING MULTI-ROLE CAPABILITY (IN THE CONTEXT OF THE LSV)

9.1 BENEFITS AND CHALLENGES OF MULTI-ROLE VESSELS

Multi-role ships offer the advantage that numerous capabilities can be provided on a single platform which offers considerable cost savings over numerous role-specific vessels. However, the particular required capability has to be available and in the right location when it is needed. A single ship being in the right location when a capability is needed at short notice is a challenge associated with putting more capabilities into fewer hulls but is an operational challenge rather than a ship design challenge. Also, if the capability is needed at short notice, the ship needs to be able to be readily adaptable to the new role. This will be hindered if modules must be transported to the ship from another part of the world. In order to negate this risk, HNoMS Maud can retain all the physical components of her capabilities embarked on board concurrently (although it is not possible for the ship to conduct all roles simultaneously). The ship's large holds and the containers allow inactive equipment to be carried on board in the event that a re-role is necessary during a deployment.

9.2 LSV CAPABILITY GROWTH

As discussed, HNoMS Maud introduces several new capabilities to the Norwegian forces and a significant step-up in capability in other areas. It is inevitable that some roles will develop/evolve while the ship is in service and she may be tasked for new/additional functions. Fortunately, she is a large platform with plenty of space and margin available in her systems to be upgraded through her life. Potential, future roles she could fulfil include:

9.2(a) Littoral Operations Support

The LSV could be adapted easily for a role supporting littoral operations providing forward positioning of troops, light vehicles, boats and supplies. She is already designed to support the smaller NorTG platforms and her deck is intended to freight CB90 combat support boats and RHIBs so could be used to host special operations deployment. Her accommodation could be modified to support larger numbers of embarked forces and her stores/ magazines could accommodate their munitions and vehicle/vessel spares. The ship's aviation facilities would prove invaluable in this role.

9.2(b) UXV Host

UAVs - The LSV's large flight deck and hangar would enable it to host Unmanned Air Vehicles, even with one NH-90 helicopters embarked. Support equipment could be modularised into containers carried on the cargo deck.

USVs & AUVs - The LSV's upper deck is designed to embark boats for logistics supply purposes and is furnished with a large heave compensated crane and two smaller support cranes. Therefore, USVs and AUVs could be embarked on deck and secured using the flexible lashing arrangements provided for TEUs and vehicles.

Alternatively, AUVs could be stowed in the holds/magazine. Their deployment and recovery could be by main crane, RHIBs or the service cranes.

UXV C2 Facilities - The ship's flexible Operations Room or large Bridge could accommodate the UXV control consoles. Several of the TEU positions are provided with multiple service connections so could embark UXV C2 modules.

9.2(c) Harbour Protection

For a harbour protection role, the LSV is able to enter and self-berth in a harbour that has been secured. The ship's own sensors and limited self-defence systems – four Kongsberg Sea Protector guns can be used to protect the ship and the harbour to some degree and the ship can act as the command node. The LSV can deploy various mission modules ashore using her deck crane, which may contain additional surveillance nodes, protection systems (barriers, lethal and non-lethal deterrents), vehicles etc. Boats including USVs and AUVs for surveillance, patrol and protection can be launched and recovered from the LSV or offloaded with their mission packages. Similarly, aerial surveillance can be provided from the LSV flight deck including UAV launch and recovery or modules deployed ashore.

9.2(d) Counter Piracy & Policing

In a counter piracy role the LSV can act as a command ship and mothership to deployed assets providing wide area surveillance, patrol, deterrence, interdiction and assistance duties. The LSV can embark various mission equipment to support these roles. For example her cargo deck can stow additional boats, which could be manned of unmanned, and these could be launched and recovered using the ships crane or more ideally by temporary installed davits. Mission containers can be embarked onto the cargo deck including command and control facilities, brig facilities or additional surveillance equipment. The flight deck and ship's helicopters will also contribute significantly to this role and UAVs could be operated from the stable LSV platform performing various duties including surveillance and deterrence.

9.2(e) HADR

The LSV provides an ideal platform for conducing HADR missions. Her large size with significant stowage capacity within holds and on deck allows the transport and offloading ashore of large quantities of supplies and equipment (including vehicles and boats). Offload can be using the ship's crane or side ramp if shore infrastructure is available. The ship can operate as command centre in this mission with additional command facilities embarked (which may be in containers). The ship's helicopter facilities are a great asset for this role and UAVs could be operated from the platform in addition to conventional helicopters.

10. CONCLUSIONS

Some question whether navies are paying a premium for flexible, multi-role ships to obtain capabilities which they only use rarely or even not at all. In the case of HNoMS Maud, it remains to be seen how the RNoN uses her. However, the Norwegian navy is small with a declared intention to enhance its role on the world stage so the LSV provides it multiple capabilities in a single efficient platform as it does not have the manpower or budget to operate numerous vessels.

The intention was to design a ship which uses as much of its capacity most of the time through clever incorporation of flexibility in the design to enable it to fulfil each of its roles effectively and efficiently.

Through careful design of its arrangement and systems, including innovative implementation of BMT's "In the Box" and "Out of the Box" modular concepts, and selection of its machinery architecture and equipment, the ship provides a very cost-effective yet operationally effective and useable solution to each of its roles in the scenarios envisaged. This has not necessarily been the case in other vessels that have attempted to employ modular solutions. Hence, the Aegir 18NR LSV, 'HNoMS Maud' brings the RNoN significantly enhanced capabilities in a modern, safe and environmentally compliant ship which is truly a multi-role platform.

With only one platform of this type in the Norwegian Navy and the operational effect she will provide not only to RNoN but to NATO the question remains: will one LSV be enough….?

11. REFERENCES

1. Kimber A & Vik AM, 'Future Naval Tankers - Bridging The Environmental Gap - The Cost Effective Solution', World Maritime Technology Conference, 2006

2. Jones S & Irgens C (Cdr SG), 'The Aegir Logistics and Support Vessel (LSV) design for the Royal Norwegian Navy – a multi-role flexible platform', International Maritime Conference, Pacific 2015

3. Def Stan 00-101, Design Standards for Explosives Safety in MOD Ships and Submarines, Part 1 - Surface Ships, UK Ministry of Defence

4. Buckingham J, 'Hybrid drives for naval auxiliary vessels', Pacific 2013 International Maritime Conference

5. Manley D, 'The NATO drive to mission modularity', RINA Warship 2018

12. AUTHORS' BIOGRAPHIES

Alex Aitken is a Senior Manager at BMT and runs the Surface Ships Outfit & Arrangement Team. Alex was the Arrangement & Outfit Design Areal Lead for the Basic Design of the Norwegian LSV. His previous experience includes the concept design of BMT's Aegir replenishment ship, supporting the PCRF on RFA Argus and the capability assessment studies for the UK's JCTS and OMAR capabilities.

Simon Jones holds the position of Chief Naval Architect (Ship Delivery) at BMT and was the Design Manager for the Basic Design of the Norwegian LSV.

PUSHING THE LIMIT: THE ROLE-LESS MULTIROLE WARSHIP

R J Pawling and **N I C Bradbeer**, UCL, UK

SUMMARY

Designing effective multirole vessels, particularly warships, is complicated by the potential for the roles, and the mix of roles, to change. Modularity is frequently suggested as an approach to addressing this problem. However, in most extant or near-term modular surface combatants, modularity only provides a limited multi-role capability and flexibility, with much capability inherent to the core vessel, and they might be described as "conventional vessels with modular characteristics". This paper reports on a study conducted at UCL into the possibility of maximising the modularity of a warship to allow both flexibility in role and infrastructure but also the ability to customise the mix of capabilities of the vessel in multi-role configurations. The concept is derived from the overall topology of container ships and conversions that have been carried out on those vessels, with a minimum "seaframe" or framework into which combinations of capabilities can be added.

1. INTRODUCTION: MULTI-ROLE AND MODULARITY

"Multi-Role" can have several interpretations; it can refer to vessels capable of carrying out several different roles in a single mission, with no equipment change; it includes "swing role" vessels which can be rapidly reconfigured to specialise in different roles (as is the case with most modern fighter aircraft). In addition to dimensions of capability, there is that of time; both the time taken to move between roles and the timescale that requirements will change. The concept of "temporal multi-role" is one origin of the study described in this paper.

Modularisation is key to the use of the US Navy Littoral Combat Ship (LCS) in multiple roles [1] as the core "seaframe" only has a minimal set of defensive weapons and sensors, similarly with the Danish STANFLEX system [2]. Where modularity has been applied to other multi-role ships, however, it has either been in the form of design and procurement modularity (such as the MEKO concept [3] or as the inclusion of a limited capacity mission bay in otherwise conventional vessels, such as the RN Type 26 or MM "Pattugliatore Polivalente d'Altura" (PPA). These vessels have a combination of permanent systems providing a capability in each of several domains.

The use of mission bays or the LCS concept of interchangeable modules for a range of weapons theoretically supports swing role adaptability with short term, theatre or even tactical level timescales. There are however complications with such role changes, as complex naval weapons have historically required significant setting to work after installation, so the time to change roles may be highly dependent on exactly which capabilities are being added or removed.

The STANFLEX system is notable as it was developed to address the strategic timescale, allowing a Danish nay composed of patrol boats suitable for peacetime operations to be re-rolled as missile craft in the even of developing hostilities between NATO and the Warsaw pact. Similar considerations led to US Navy studies for large warships designed with extensive combat system modularity to allow system updates and role re-balancing through life, in the context of a technologically-focussed Cold War [4], [5]. Figure 1 illustrates one such "variable payload destroyer" design, based on the DDG-51.

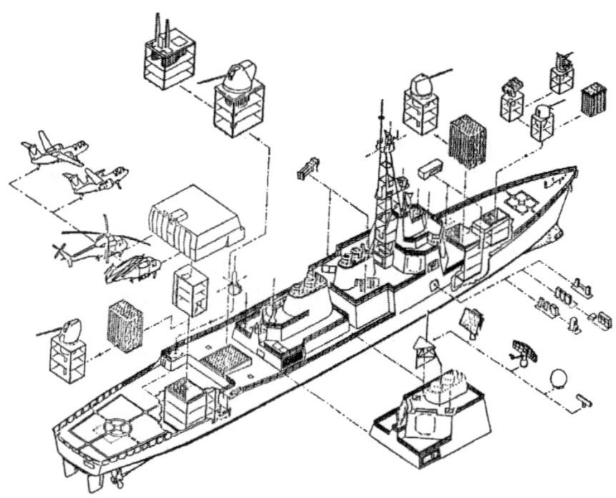

Figure 1: An example of a variable payload destroyer [5]

Notable in the US Navy studies, including the SEAMOD concept [6] was the inclusion of supporting systems and spaces, both for combat systems and the mechanical and electrical systems, in the modularisation system. Although the most visible aspects of a warships capability are its' weapons and sensors, these are underpinned by hotel systems such as HVAC and electrical generation, and the various crew support infrastructure, so a multi-role and modular ship design has to include these aspects as part of the capability.

The extreme example of the use of modularisation to achieve a change in role is perhaps the Reserve Merchant Ship Defence System (RMSDS), also known as ARAPAHO, which made use of a large number of standard intermodal containers to convert a container ship into a carrier for ASW helicopters and STOVL aircraft [7] A similar, unrealised proposal for a vessel with a swing role capability at the strategic level was the "PROTEAN" small aircraft carrier from YARD [8].

Figure 2 illustrates the concept underlying PROTEAN, where some aspects of commercial vessel design style, such as wing tanks and double hulls in container ships, were adapted to give a hangar deck with large module bays (referred to as "library bays") on either beam. These modules would house all spaces associated with the vessel's current role, such as air workshops and pilot accommodation for a carrier or troop accommodation for an LPH. The basic vessel could then be re-rolled through life.

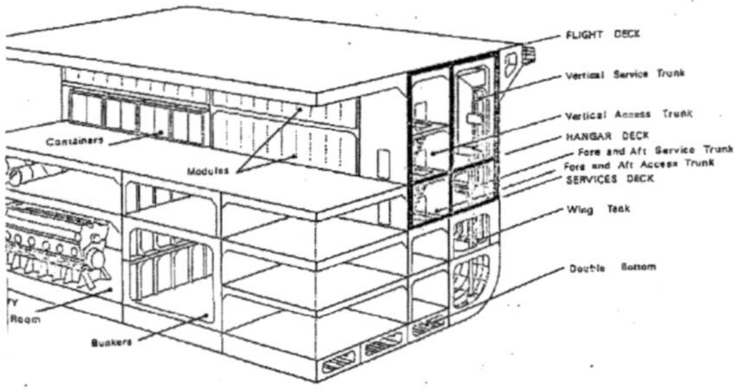

Figure 2: Typical hull section in PROTEAN [8]

2. CONSIDERATIONS IN MODULARISATION

2.1 SELECTION OF MODULE SIZE

Module size varies significantly across different systems, both in service and proposed. A STANFLEX container has a length of 3.5m, width of 3m and height of 2.5m, with a basic weight of 3.5-5 tonnes, whilst the "B" size VLS module in [5] being a 64 cell MK41 battery having a combined length of 10.2m, width of 6.8m, height of 7.7m and empty mas of 116 tonnes!

Type	Length, m	Width, m	Height, m	Mass, t
STANFLEX	3.5	3	2.5	3.5-5
MEKO				
Weapons	4	4.7	2.66	
Electronics	3-3.5	2.44	2.15	
PROTEAN				
Basic	6.096	4.064	4.064	
Large	6.096	32.512	4.064	
Bay size	6.096	32.512	5.08	
Bespoke	10	6.1	3.25	
USN VLS B	10.2	6.8	7.7	116
ISO containers				
40 foot	12.192	2.438	2.591-2.896	2.2 tare
20 foot	6.058	2.438	2.591-2.896	3.8 tare

Table 1: Some module sizes in naval ship design

Table 1 above summarises the sizes of some module systems proposed or used in warships. The selection of an appropriate module size is a trade off between several factors, including transportation, efficiency and utility, lifting weight limits and cost. The latter is difficult to address, as a perceived low risk, low cost solution (using an off-the shelf commercial module form factor) may lead to a poor arrangement, and may not represent any significant savings if the enclosure itself is only a small portion of the system cost.

2.2 TRANSPORTATION OF MODULES

A consideration in module sizing can be transportation of the modules; they may be deployed to friendly bases in an operational theatre to allow re-rolling or repair by replacement; or construction and subsequent support of a national fleet may be geographically and organisationally dispersed (an important consideration in some types of strategic and unconventional warfare).

Loading gauges are closely related to the ISO 668 container size. For road transport in the UK, an "abnormal load" is defined as being longer than 18.65m, wider than 2.9m or with a combined vehicle mass of more than 44 tonnes [9], but more typically a width limit of 2.55m applies [10]. UK enhanced rail loading gauges W9-W12 similarly have maximum container widths of 2.55-2.6m and heights of 2.74-2.9m [11]. This would tend to dictate a preference for modules around the side of ISO containers. Where pre-positioning by air freight is to be used, it should be noted that air freight modules are generally not cuboid, expect the AMA type container, which is 3.18m in length and 2.25m in width and height, with a maximum laden mass of 6.8 tonnes.

It should be borne in mind that the size of the module is not the same as the module "pitch" on the vessel. Taking the case of ISO intermodal 20/40 foot containers; they are attached to the deck and to each other by twistlocks at each corner. These impose a vertical spacing of between 20-50mm between modules. Similarly, there will be some need for space to connect systems connections, arrange access etc. Thus, a modular vessel will have a module pitch; i.e. space allowance; larger than the module size.

2.3 USEFULNESS OF A MODULE

The ISO 668 standard intermodal container is an attractive starting point for "modular thinking" because many people are aware of modular housing and equipment used in commercial land-based applications. One of the first tasks in the design study described here was to conduct a survey of containerised equipment, facilities and accommodation to gain a better understanding of some of the detail aspects.

Diesel generators up to approximately 2MW [12], reverse osmosis plants [13] and chilled water plants [14] are all available in 20- or 40- foot container dimensions for land based civil applications. However, the situation is more complicated in the case of accommodation and working spaces, rather than equipment enclosures. Width is frequently greater than the ISO standard, at 10ft (3.048m) [15] to allow access past beds, furniture etc, whilst the internal height of a standard ISO container can rapidly be used up with flooring and light fixtures so heights can range from the 9.5ft "high cube" to 3.3+m for A60 rated modules with the associated fire protection linings [16].

It is important to consider access when modularisation is used. Modularised equipment for land use will frequently be in a location with ample space for maintenance, or may simply be returned to the supplier, so designed-in maintenance space may be highly restricted. Arranging offices and particularly accommodation with access at the ends (as in a cargo container) is difficult and land-based applications frequently have the doors in the longer sides. Configurations are also available where the entire side of the container opens, or is removable, however the largest size of opening will always be limited by the need for corner pillars to support a stack.

2.4 EFFICIENCY OF MODULARISATION

2.4 (a) Space

As a modularisation is applying a discrete variable (number of modules) to a largely continuous space (size of ships, their systems and spaces), the packing efficiency will vary with ship and module size. Figure 3 below illustrates the changes in tonnes displacement per module and fraction of internal volume used for a generic frigate-type hullform completely packed with 40-foot equivalent modules.

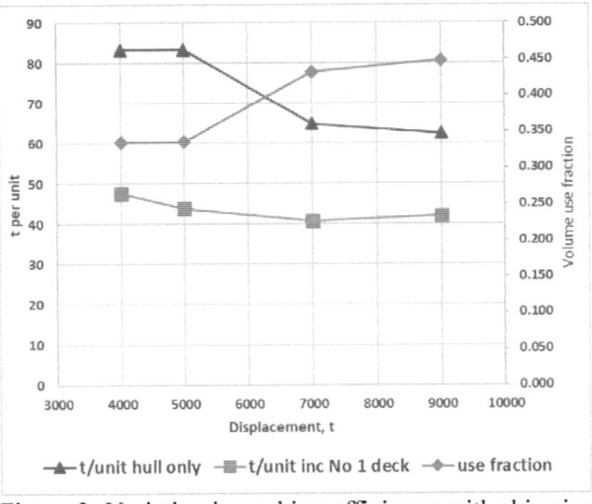

Figure 3: Variation in packing efficiency with ship size

This figure clearly shows a step in efficiency as the hull grows to accommodate a complete row of modules. Including upperdeck modules, however, leads to a more constant packing fraction. This illustrates the impact of the fine ends of a warship hull.

Figure 4 shows a generic set of module locations, arranged into "holds" similar to a container ship, separated by double bulkheads.

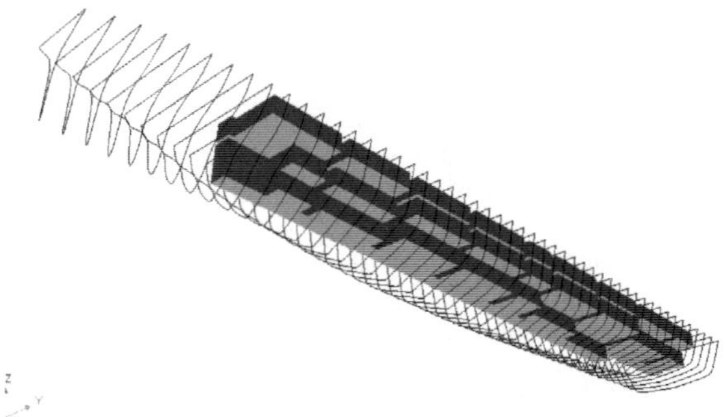

Figure 4: One possible arrangement of modules into "holds" in a frigate-type hull

Figure 5a below shows the distribution of modules required to enclose individual Building Blocks in a baseline conventional 6000t frigate design, with Figure 5b showing the division between Functional Groups.

Figure 5a shows that 40% of the Design Building Blocks in the baseline model require at most one short (6m) module, and that many items do not fit an integer number of modules efficiently. Figure 5b provides some context to this; many of the smaller items are small electronics spaces for local control of combat systems (FIGHT) and various items of hotel equipment (INFRA). Although the hotel systems could be grouped into the AMR modules described above, the overall impact is that the majority of small spaces will be larger than required in the modular ship.

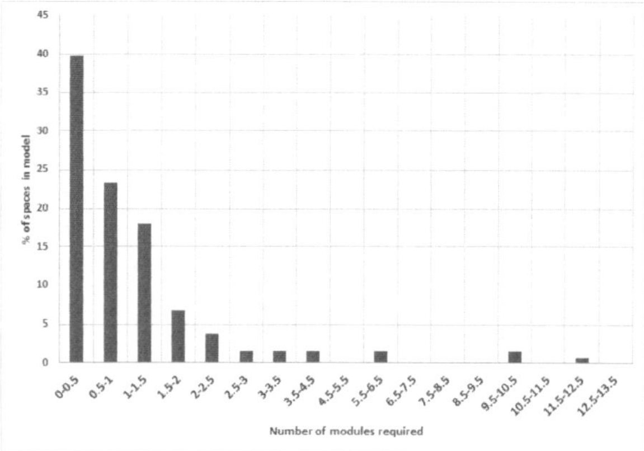

Figure 5a: Distribution of module size equivalencies for a 6000t multi role frigate

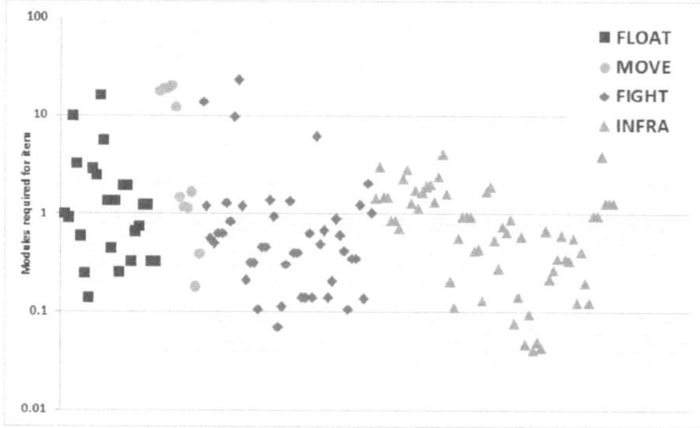

Figure 5b: Individual Design Building Blocks and their module equivalencies for the 6000t frigate

2.4 (b) Weight

Empty ISO containers typically have a mass of 2-4 tonnes and can support significant stack loads. For example, a commercial 20-foot spaceframe module supporting 96 tonnes of stack weight is available with a self-mass of 4 tonnes [17]. However this equates to a 7mm thick steel box of equal dimensions and non-structural steel partition panels are typically less than 2mm thick, so a 20-foot unit not intended to support any stack load would have a mass of 1.2 tonnes. Containerised housing systems may use much lighter structures, meeting the ISO standards in dimensions only.

The purpose of this simplistic comparison is to illustrate that using commercial container enclosures may make modules excessively heavy, unless stacking is actually required.

2.5 ARRANGING A MODULAR SHIP

The generation and analysis of ship general arrangements is a current area of research and development, as described by Andrews et al [18]. Typical influences on arrangements include; personnel flow; noise and vibration; motions; and survivability. A reconfigurable, modularised warship adds some areas of additional complexity.

2.5(a) Likelihood of Change

A space or system that is more likely to be changed should be in a location requiring less module moves to carry out the change.

2.5(b) Complexity

A space or system that is composed of multiple modules or has direct connections with adjacent modules may be more complex in a modular form than a conventional one.

2.5(c) Vertical vs. Horizontal Connections

The use of extensive modularisation will tend towards a preference for vertical connections between modules and horizontal connections through permanent dedicated systems runs.

2.5(d) Difficulty of Removal

Module weights must be considered with respect to available craneage for conversion operations. Modules should be arranged so that difficult to move modules are not in the removal route of modules likely to be changed. Modules that are extremely unlikely, or impractical to be changed, should be concentrated.

In contrast with these complications, the actual process of developing the arrangement was more straightforward for the modular than a conventional ship. With a reduced number of possible increments in dimensions for a space, arrangement was reduced to a space allocation problem rather than an allocation-and-sizing one.

3. AN EXAMPLE MODULARISED DESIGN

This section describes a study into the application of extensive modularity to a baseline multi-role surface combatant design, to develop a design that could be extensively re-rolled and re-equipped through life. The design was developed using the architecturally-centred Design Building Block approach developed at UCL and previously used in a range of studies including uninhabited vehicles [19] and electric weapons [20].

3.1 OVERALL DESCRIPTION

To examine possible impacts of adopting a highly modularised warship design approach, an existing UCL design developed in 2012 was modified. This baseline design is shown in Figure 6a and is a 6000 tonne vessel with capabilities broadly similar to the Type 26 as presented to the public at that time, but using the UCL Ship Design Exercise Database. This design was developed as a "straw man" design to be used in future studies.

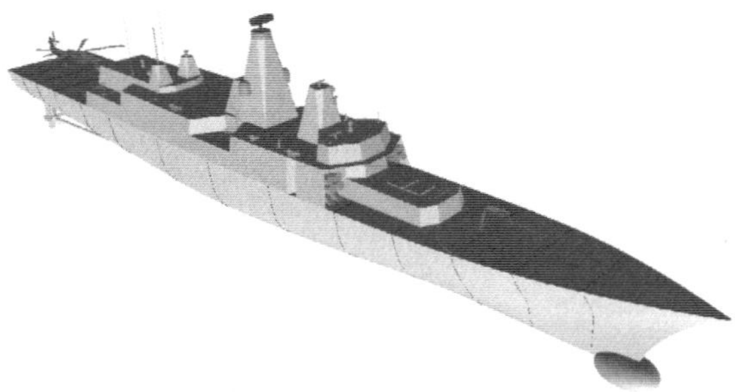

Figure 6a The conventional 6000 tonne UCL design

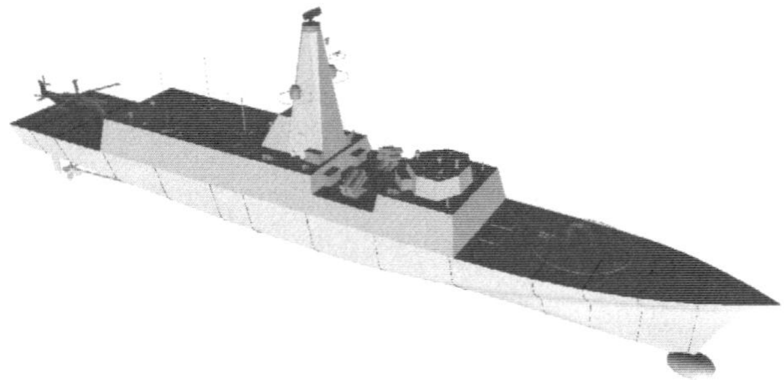

Figure 6b: The modular 7000 tonne UCL design

Figure 6b illustrates the first version (V1) of the modularised design, with its' principal particulars given in Table 2.

Table 2: Principal particulars of the 7000 tonne design

Disp		Length, WL	142.3m
Light, BoL	5900t	Length, OA	148.3m
Deep, BoL	6860t	Beam, WL	19.3m
Light, EoL	6260t	Beam OA	21.9m
Deep, EoL	7210t	Draught	6.2m
Enclosed Volume (V1)	33400m^3	Depth, amidships	13m
Overall density (deep)	0.216t/m^3		
Complement	130	Endurance Speed	12 knots
Accom.	166	Range	8000nm
		Stores	60 days
AAW	48 x Sea Ceptor, decoys	Strike	16 x Tomahawk, 127mm with 300 rounds
ASW	2 x triple tubes, air weapons	ASVW	8 x RBS15 Mk3, 2 x 30mm
Radar	Artisan	Sonar	2050, 2087
EO	IRST	Aviation	1 x Merlin
L&R	Side davits	UXVs	2-4 11m USVs

Considering the discussions above, the module "pitch" used in the ship design was 6.5/12.5m in length, 2.6m in width and 2.9m in height, with service connections assumed to be in the ends and access in the long sides of 12m modules and ends of 6m modules. Deck heights were a minimum of 3.1m to allow for connections.

3.2 HULLFORM AND POWERING

The hullform developed for the example modularised warship design was influenced by the Deep Vee hullform proposed for use in frigates and destroyers [21], and is illustrated by the sections plan shown in Figure 7. The hullform is characterised by a low Cm (0.67), high Cw (0.79) and a long parallel mid-body (83m of parallel upperdeck, reducing to 8m at the keel). This hullform style was chosen to increase area on Nos 1, 2 and 3 decks.

Initial resistance and powering analysis was carried out using Holtrop and Fung 2 methods in Paramarine, including a comparison with a "conventional" 7000 frigate hull. The resistance estimation was carried out in the deep condition and included 6 months of tropical fouling and waves equivalent to sea state 3. The results of this analysis are summarised in Figure 8, and, somewhat unhelpfully, show that the vee-shaped hull may be better, worse or the same as the conventional hull.

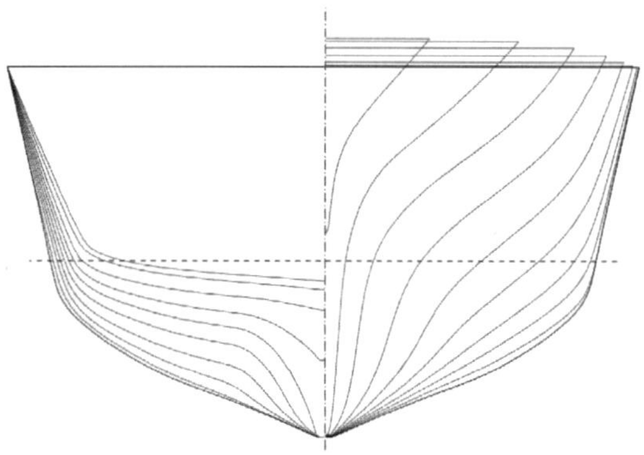

Figure 7: Sections plan for the example hull

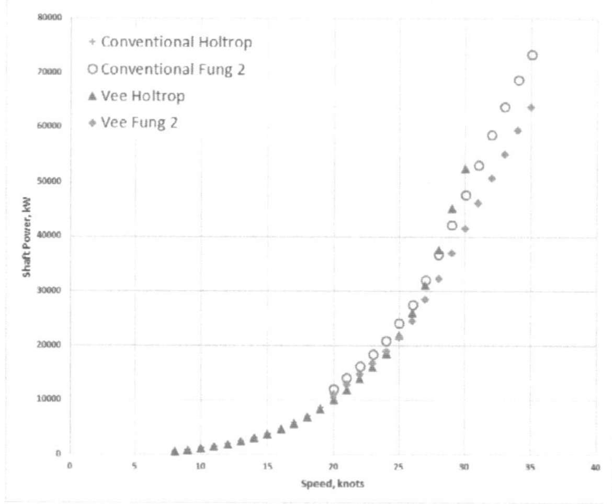

Figure 8: Comparison of shaft power estimates for conventional and vee shaped hullforms

A COmbined Diesel eLectric And Gas turbine (CODLAG) propulsion architecture was selected, as in the conventional basis design. On a single MT-30 driving both shafts through a splitter reduction gearbox the design could make 27-28 knots. Adding the two cruise 3.5MW shaft motors increases this to 28-29 knots. The use of a stern flap adds approximately a knot to maximum speeds.

3.3 THE CORE SHIP

The equipment and spaces identified as being permanent "core ship" were concentrated into as few vertical sections as possible. The core ship in this case consisted of:

- Mooring spaces forward.
- Forward pump room
- Fuel tanks
- GT machinery room and GT ducting*
- GB machinery room
- Motor room and motor controllers
- Shafting and steering gear
- Ship control centre (SCC) and engineering office*
- Workshops and engineering stores*
- Ships boats*
- Mainmast*

This is illustrated in Figure 9, which shows how the items marked with an asterisk were grouped into a single section roughly amidships.

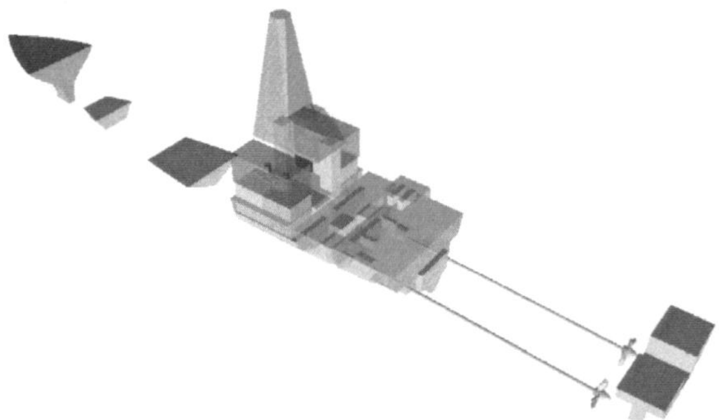

Figure 9: The "core ship" items

3.4 GENERAL ARRANGEMENT

Figure 10 shows the overall configuration of the modular design, with all Design Building Blocks visible. As in previous UCL DBB design studies, the colour scheme is; Blue – FIGHT Functional Group; Yellow – MOVE Functional Group; Red - FIGHT Functional Group and Green – INFRASTRUCTURE Functional Group.

Figure 10: Overall model with all Functional Groups visible

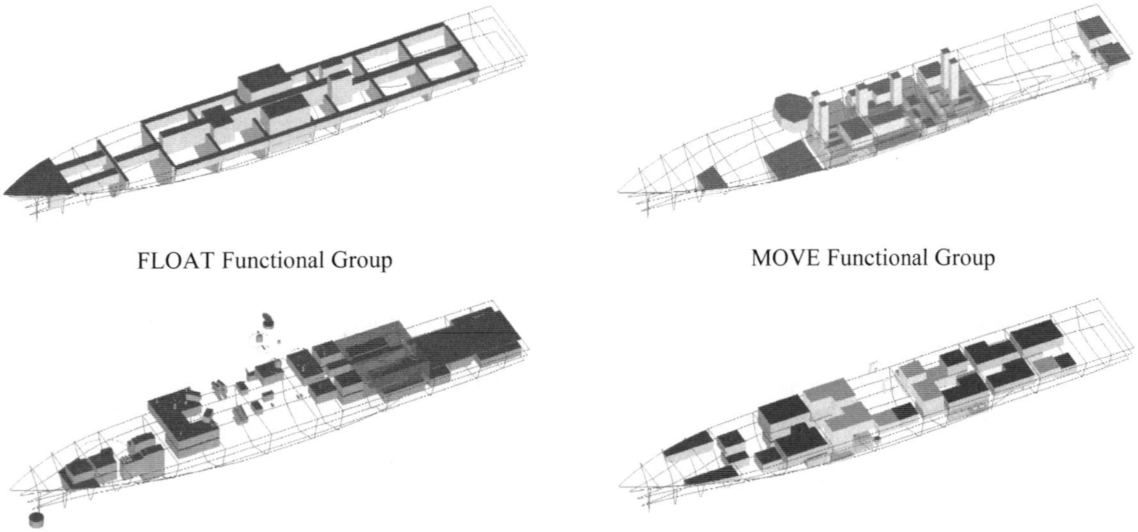

FLOAT Functional Group — MOVE Functional Group

FIGHT Functional Group — INFRASTRUCTURE Functional Group

Figure 11: Design Building Blocks of each Functional Group

The four images in Figure 11 show each of the four Functional Groups in the modular design. The overall arrangement is relatively conventional. FLOAT highlights the use of a longitudinal systems run on either beam and the main access on the centreline. The access routes were also sized using the module pitch, so that the space could be reconfigured if required. The watertight compartments are separated by double bulkheads, with an assumed uniform depth of 1.5m, it being expected that the actual depth would alternate between those designed to offer vertical access, and those needed for structural strength or systems interconnections.

Consideration of the influences on arrangement described in Section 2.5 led to the command and electronics spaces being concentrated in the forward superstructure under the bridge, as they were designated as having a high likelihood of change during life. Ballistic protection was added to this part of the superstructure and the Ship Control Centre was located aft to provide some degree of separation in ship and damage control functions.

Figure 12 shows a cross section of the ship in way of the forward superstructure, including the centreline passageways and access lobbies, officer's accommodation blocks and stores in the lower hull. Also visible is the large amount of void volume that results in the forward parts of the hull.

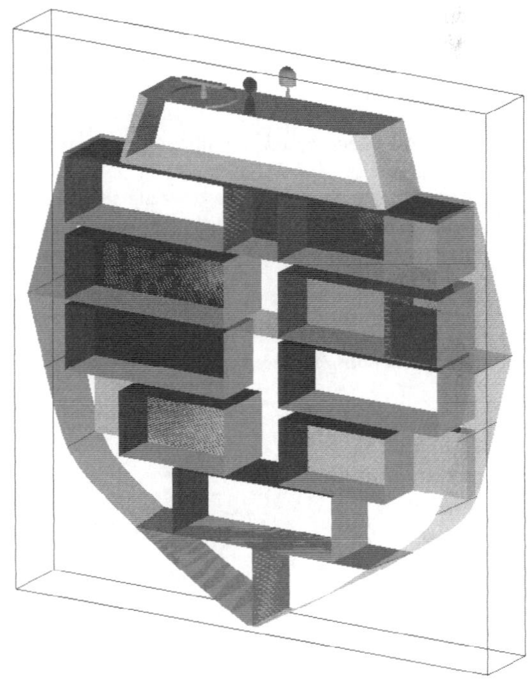

Figure 12: Section in way of the forward superstructure

Propulsion, generation and auxiliary machinery was designated a having similar liklihoods of requiring major change and similar levels of difficulty, so was concentrated. Auxiliary machinery was located in modules on the mezzanine deck of the machinery spaces, on either beam, as shown in Figure 13 below.

The otherwise unusable space in the bilges was used to form wing tanks for fuel and other liquids storage. The modular design has three main options for liquids storage; in the deep double bottom; in wing tanks; or in the lower part of the double bulkheads. Each has its' own merits and this study only assessed the high-level design impact; asymmetric flooding, but the resulting angle of heel was not found to be significant.

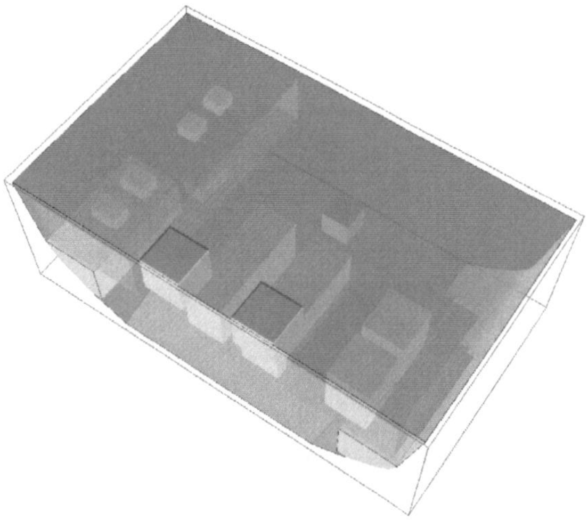

Figure 13: Forward machinery space

3.5 STRUCTURES

As noted above, a concern in a ship making extensive use of internal modules is the reduction in longitudinal strength from large openings in the upperdeck. In initial sizing, an increased overall structural density based on commercial vessels was used, of 90kg/m^3, resulting in an estimated mass of 2700tonnes for the hull structure. The midships section was examined in further detail and a distribution of equivalent thicknesses developed based on the longitudinal bending moment, shown in Figure 14.

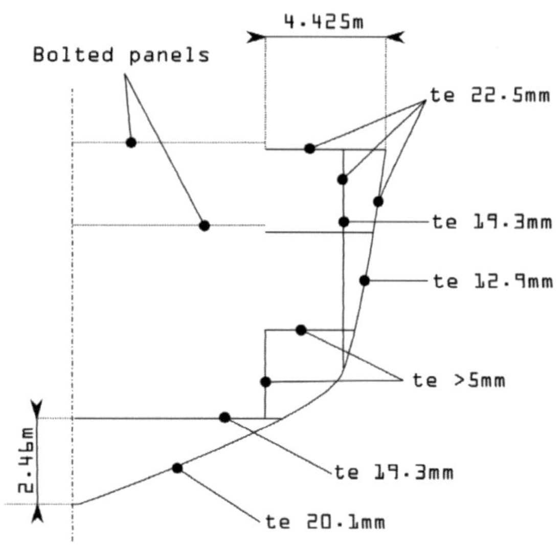

Figure 14: The selected equivalent thicknesses for the midships structure

The upperdeck stress was found to be the dominant aspect of the design, and the original concept of a purely box-girder based structure was found to lead to impractically thick plating. Nos 1 and 2 deck were extended inboard until equivalent thicknesses reduced to those comparable with the NFR-90 midships structure described by [22]. The effectiveness of the box girder systems runs and the deep draft helped limit the required "overhang" to a single modules' width (2.6m).

Also shown in Figure 14 is the very low thickness of plating required in the bilge area, which provides a seat for the modules on No 3 deck. This is largely ineffective in longitudinal bending, but would instead be dictated by the need to integrate the heavy keel and upper side structures. It is not expected to be 5mm thick, but this is the minimum for longitudinal strength.

Using this midships section the hull primary structural weight was estimated as 2290 tonnes. The more pessimistic initial estimate was thus retained to account for the larger mass of secondary and detail structure required, including removable panels (assumed non effective in longitudinal bending) on Nos 1 and 2 deck.

3.6 WEIGHTS AND STABILITY

3.6(a) Weight Breakdown

Figure 15 shows the breakdown in weights as a percentage of the deep ship, end of life displacement. The structural weight, at 46% deep or 54% light weight, is higher than historical data for surface combatants (40-50% lightweight, [23] but may not be so notable compared to modern vessel structural styles.

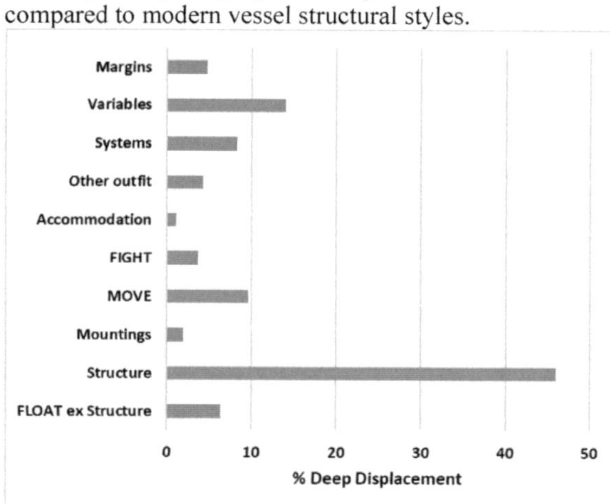

Figure 15: Weight breakdown for deep displacement

The overall density of the design reduced, however, from 0.246 to 0.215. The latter change can be explained by examining the allocation of spaces and equipment to modules – as shown in Figures 5a and 5b, many of the systems and spaces in the baseline 6000t design require fractional modules. Rounding this up to integer numbers of modules, and introducing larger box girders, will clearly increase the internal volume of the design.

3.6(b) Intact and Damaged Stability

The design was assessed against DEFSTAN 02-109 GZ shape requirements in deep, operating and light load conditions at beginning and end of life (i.e. including growth margins). It passed all criteria in each case, with two notable issues; firstly, trim by the stern was excessive in the light load conditions (1.3m); and secondly the GMtf was higher than desirable in all loading conditions, being highest (2.8m) in the deep condition at start of life. The latter was driven by the full waterplane, and indicates further refinement of the design is necessary.

Figure 16 shows the waterlines generated from a range of 3 compartment damage cases, and 4 compartment cases at the extreme ends. The ship passed DEFSTAN 02-109 in all cases, with the upperdeck remaining above the waterline. However, the standardisation of compartment length at 7.5m and 13.5m between bulkhead centres leads to a large loss of buoyancy and lower damaged freeboards than might result in a conventional vessel with a wider range of compartment lengths.

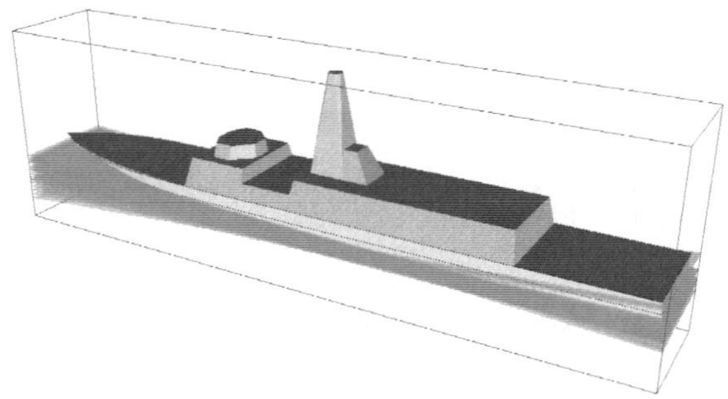

Figure 16: Set of damaged waterlines

4. VARIATIONS ON THE DESIGN

To complete this study, examples of through-life capability and equipment changes have been examined. As of the time of writing this paper, this part of the study is still underway and will be more detailed in the conference presentation.

4.1 Warfare Mode Switch

An example of a major change in role that is of interest is the possibility of a change in strategic situation requiring a shift from mission bays and UXVs to increased kinetic capabilities. This was implemented by doubling the Sea Ceptor and cruise missile batteries, and replacing the UXV facilities with 12 conjectural hypersonic missiles.

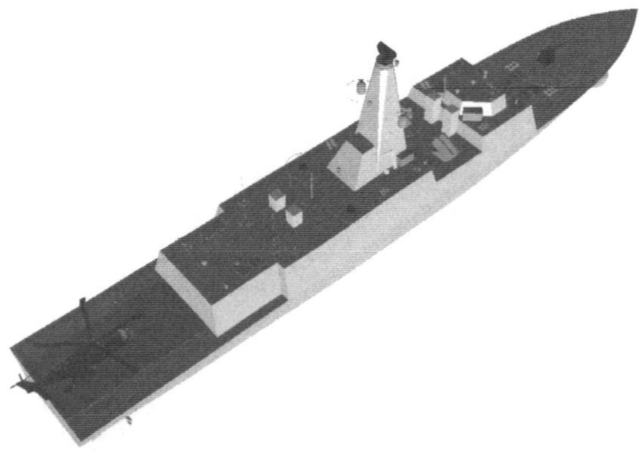

Figure 17: A variant design with increased kinetic capabilities

Designing in allowances for later increases in VLS size is not unusual, being included on RN Type 45, Dutch LCF amongst others, however this space will frequently find another use early in life. Figure 18 shows a detail of the arrangement in way of the hangar and hypersonic missile launchers.

Figure 18: Changed aft arrangements accommodating large missile launchers

Although the modular design allowed workshop and accommodation modules to be rearranged, the key advantage was the potential to reconfigure areas deep in the ship; UXVs generally demand upperdeck space and superstructure volume; whilst missiles required hull volume and can be demanding in terms of height.

This specific variant raises an additional question regarding the integration of magazine spaces, including VLS batteries. Magazine spaces will require enclosures to more demanding standards, and connections to higher capacity water systems for firefighting. The latter consideration drove the inclusion of a dedicated pump room forward in the "core ship".

5. DISCUSSIONS

5.1 SURVIVABILITY

The adoption of several features more typically seen in commercial vessels raises questions about survivability, and these are considered within the standard definition of survivability.

5.1(a) Susceptibility

The angled superstructure side panels shown in Figure 12 prevent any impact on Radar Cross Section (RCS) from exposed containers, and the higher GMt of the vee-shaped hullform would enable fitting of larger infra-red signature reduction measures for exhausts. Underwater acoustic signature is more complicated; the larger machinery spaces may make silencing easier by permitting larger enclosures, and upperdeck machinery could be used. There is the possibility of additional noise generation from containers vibrating or rattling against their mounts, although this is not known to be a serious problem on cargo ships. It should also be noted that a modular ship system implies more interfaces between different materials and structures, which will usually increase vibration attenuation.

5.1(b) Vulnerability

The use of double bulkheads, needed to provide resistance to torsion, is likely to limit the spread of blast and fragmentation damage. The enclosure of equipment in containerised modules will also provide additional protection from fragments and from moderate blast overpressures. Modularisation provides more options for shock protection of equipment, as it can be mounted at the module or equipment level, but this must be borne in mind when considering module height as additional space is required for mounts and equipment motion. Response of the modularised systems to underwater shock may be challenging to design for, in particular captivity requirements. Commercial twistlocks for securing ISO containers have both cast and forged parts, with very high compressive strength, but the primary concern for shock loading will be the cone-shaped locking elements failing under tension at the "top" of a shock response.

Where the use of a module system leads to longer watertight compartments this will lead to lower quasi-static pressures (QSP) from internal explosions. The blast and shock performance of the bolted panels used on Nos 1 and 2 deck is unknown; bolted hatches are used for removal routes, but in this case they have a much larger area to perimeter ratio. However, as conventional decks are likely to vent in an internal explosion, the nett impact may not be significant.

The structural section shown in Figure 14 requires more analysis of response to damage. Although box girders can provide significant residual strength, in this design the loss of a shear strake or bilge due to a collision, for example, will remove much longitudinal structure. Another concern is how the unsupported inner edges and light bilge structure will perform under shock.

5.1(c) Recoverability

The sided systems runs provide good separation of duplicated supply systems, reducing vulnerability, but may complicate the issue of restoring supply runs after damage, when compared to a conventional passageway arrangement. The use of deep double bulkheads as connection points for longitudinal systems may offset this.

Although modularisation would make repair by replacement more straightforward, the possibility of distortion of modules in a damaged compartment may complicate damage control efforts. The historical use of ISO containers to construct targets for weapons tests may provide some information on this aspect.

5.2 THE IMPACT OF MODULARISATION

Moving to a modularised design increased displacement by almost exactly 1000t (deep EoL for the original design being 6260t), length WL by 3m, draught by 0.9m and beam by 2.2m, although the overall density of the design reduced from 0.246 to 0.215 due to additional void or unused space required by extreme modularisation. The majority of the increase in mass was in the primary structure, but the main longitudinal system runs also increased in mass, and installing sufficient systems to support a range of possible future requirements and equipment may be a more significant impact on an increase in cost. Against this are potential reductions in initial procurement cost due to improve producibility from modular working, as it significantly reduces the amount if fitting-out carried out in the ship and potentially opens a wider range of equipment suppliers.

6. CONCLUSIONS AND FUTURE WORK

The design study described in this paper was intended as a "what if" look at the extreme limit of the established concept of modularisation in warship design; to design a frigate that was essentially a small container ship. The main outcome is that the application of an "extreme modularisation" approach to the design of a 6-7000 tonne general purpose surface combatant appears to be possible, in that the main areas of concern (module sizing; use of internal volume; viability of an open-hold type structure) do not by themselves rule out the concept.

The study began with a survey of existing modular equipment and this can be summarised as; *all containers are equal, but some are more equal than others* – within the overall concept of the ISO container, there are many variations, and this may mean that the conventional corrugated steel box may not be the best option.

There are several areas of future work; at the time of writing variant studies and costing estimation are still underway. On the analytical side, FEA analysis of shock response of both the structural arrangement and a fully modularised internal arrangement are of interest for future work.

7. REFERENCES

[1] Morral, D., "Littoral Combat Ship (LCS) Mission Modules", 2004

[2] Grooss, P. & Stihoj Pedersen, A., "Naval Requirements: The Danish Perspective"

[3] Wurm, H-G., "Impact of Modularity Through Function Units on Refit and Modernisation", International Naval Technology Expo, 1982

[4] Broome, G.W., Nelson, D.W. & Tootle, W.D., "The Design of Variable Payload Ships", Naval Engineers Journal, April 1982. ASNE.

[5] Boerum, R.R. & Birindelli, J.B., "How Modular Combat Systems will Enhance Support of Surface Combatants", Naval Engineers Journal, November 1985. ASNE.

[6] Lawson, C.E., "SEAMOD – A new way to design, construct, modernize and convert US Navy combatant ships", 14 Annual Technical Symposium, ASE, 1977

[7] Basile, N.K. & Mashin, N.P., "The Reserve Merchant Ship Defence System (RMSDS)", Naval Engineers Journal, April 1977. ASNE.

[8] "PROTEAN", YARD Ltd Consulting Engineers, 1975

[9] https://www.gov.uk/esdal-and-abnormal-loads

[10] https://www.gov.uk/guidance/moving-goods-by-road

[11] http://www.freightonrail.org.uk/HotTopicsTenProposedEnhancementsScotland.htm

[12] https://powercontinuity.co.uk/containerised-generators/

[13] https://www.eurowater.com/products/water_treatment_in_container.aspx

[14] https://www.kti-plersch.com/products/refrigeration-systems/chiller/ccwp/

[15] https://www.mac-containers.co.uk/sale-hire.htm

[16] https://www.hooverferguson.com/products/ccus.html

[17] https://cargostore.com/20ft-dnv-container-frame-skid/

[18] Andrews, D.J., Duchateau, E.A.E., Gillespe, J., Hopman, J.J., Pawling, R.J., Singer, D.J., "State of the art report: Design for layout", International Marine Design Conference (IMDC) 2012, Glasgow.

[19] Pawling, R.J., Andrews, D.J., "Large Uninhabited Vehicles and the Minor War Vessel", International Conference Warship 2013: Minor Warships, RINA, Bath, June 2013.

[20] Andrews D.J., Bucknall, R., Pawling, R.J., Greig, A., McDonald, T., "The Impact of Integrated Electric Weapons on Future Warship Design Using Conventional and Unconventional Hullforms", Engine as a Weapon (EAAW) 2011, IMarEST, London.

[21] Kehoe, J.W., Brower, K. & Serter, E., "The Deep-Vee Hull Form – Improves Seakeeping and Combat System Performance", Naval Engineers Journal, May 1987. ASNE.

[22] Schaffer, R.L. & Kloehn, H.G., "Design of the NFR-90", Naval Engineers Journal, March 1991. ASNE.

[23] Kehoe, J.W. Graham, C. Brower, K.S. & Meier, H.A. 1980b. Comparative Naval Architecture Analysis of NATO and Soviet Frigates – Part II. Naval Engineers Journal. December 1980. ASNE.

8. AUTHORS BIOGRAPHY

Dr. Rachel Pawling is a Lecturer in Ship Design at UCL, teaching the subject of ship design to undergraduate and postgraduate students. She obtained her PhD in computer aided ship design in 2007 and has subsequently continued her research in projects funded by the EC, UK and US governments.

Dr. Nick Bradbeer is a Lecturer in Naval Architecture at UCL, teaching both surface ship and submarine design. He obtained an MSc in Naval Architecture from the department in 2003, then returned to complete his PhD in structural survivability. His research interests include operational modelling and analysis and applications of VR in teaching and design.

THE MULTI-ROLE SUPPORT SHIP FOR HUMANITARIAN ASSISTANCE AND DISASTER RELIEF OPERATIONS

D Buxton, Wade Carson, VARD Marine Inc., Canada

SUMMARY

The paper examines the contemporary multi-role platform as an increasingly sought-after national capability and discusses how this ship-type, if designed and outfitted appropriately, can provide a range of capabilities to address the constantly evolving nature of naval doctrine and maritime operations. The motivations that drive acquisition decision makers to consider this type of capability are examined, namely: operational flexibility, doctrinal relevance and affordability. An examination of the pitfalls of acquiring a multi-role platform is also presented and discussed. An overview of how modern design techniques can be applied to ensure the ship owner is provided a range of options that are easily tailored to their unique combination of requirements follows. The paper concludes by presenting a typical modern multi-role vessel within the context of a humanitarian assistance and disaster relief scenario to illustrate the utility of these vessels as an effective response capability in a non-combatant role.

NOMENCLATURE

HADR	Humanitarian Assistance and Disaster Relief
MRSS	Multi-Role Support Ship
MRV	Multi-Role Vessel

1. INTRODUCTION

What is a multi-role vessel? In the strictest sense any ship which can perform two or more functions could be identified as a multi-role vessel. Indeed, there are many diverse examples of ships that are described or labelled as being multi-role platforms. It is therefore true that the term multi-role has many connotations and can mean different things to different people. In fact, there are ships that operate at each end of the spectrum of naval operations, from logistics and combat support to high-intensity combat operations, and at many points in between, that use the moniker of *multi-role vessel*.

Looking at the Canadian example, we see at one end of the spectrum the recently announced design for the Canadian Surface Combatant program, based upon the Royal Navy's Type 26 Global Combat Ship. This ship has been described as a multi-role frigate with a centralised mission bay and that can be adapted to employ a range of weapons and sensors to allow it to execute different roles as the strategic situation requires.[1] At the opposite end of the capability spectrum there is MV Asterix, a commercial container ship converted into a multi-purpose naval support ship able to offer a range of capabilities, including: underway liquid and solid replenishment, humanitarian assistance and disaster relief, fleet medical support, helicopter maintenance, and, command and control facilities.[2]

There are many other diverse and wide-ranging examples such as the USN Independence-class littoral combat ship, HMNZS Canterbury, a multi-role vessel in service with the Royal New Zealand Navy (RNZN), the Royal Danish Navy's StanFlex frigates and, the Italian Navy's newly launched Multipurpose Amphibious vessel (LHD), to name but a few.

For the purposes of this paper, however, the notion of what constitutes a multi-role ship is deliberately more focussed. This paper will concentrate on the naval auxiliary or support ship and not ships intended for high-intensity combat operations. The *multi-role support ship* (MRSS) is defined generically as a vessel that competently performs a combination of support functions typically delivered by naval auxiliary and/or amphibious ship types. That suite of capabilities could include fleet replenishment, logistical resupply and tactical sealift, logistics over the shore, sea basing and afloat maintenance support, deployable command and control facilities and/or enhanced medical services.

The nature of a typical MRSS will allow governments to use these vessels to satisfy military, para-military and non-military (or civil) mission requirements. As such, they also serve as effective tools for foreign policy to further international relations and diplomacy in response to peace support operations in failed states, to tackle international criminal activity such as piracy and human trafficking, and to provide relief from natural disasters and address humanitarian crises.

2. MARITIME OPERATIONS AND THE EMERGENCE OF THE MRSS

While the context within which any given naval force operates will vary, based upon regional factors, political influences and doctrinal differences, the one constant is uncertainty. The operational context in the last 30 years has constantly evolved and perhaps remains both in a constant state of flux and as varied today as it ever has been. The uncertainty regarding what the next deployment might involve presents a challenge for procurement organizations, military planners, and operational authorities alike.

Following the cold-war era, we have seen less emphasis on traditional war-fighting force structures necessary in a bi-polar world order. The emergence of failed states and non-state actors, international terrorism and trans-border crime have produced both asymmetrical threats and humanitarian crises of significant proportions.

Natural disasters and extreme weather events, such as droughts, floods, and hurricanes, also continue to create dire situations of immense human suffering. Hurricane activity, for example, has been steadily increasing in terms of strength, intensity, frequency and duration, since the early 1980s.[3]

Accordingly, navies have had to adapt and become better equipped and trained to deal with military operations other than war. As depicted in Figure 1 below, the suite of capabilities that would be delivered by the MRSS is situated on the left-hand side of the conflict continuum, ranging from crisis response to combat support operations. While the ability to operate in contested environments can be varied for any given ship design, depending upon the addition of self-defence capabilities and off-board weapon systems, it is expected that the vessel will not normally be placed in harms way without the protection of a task group or a warship consort.

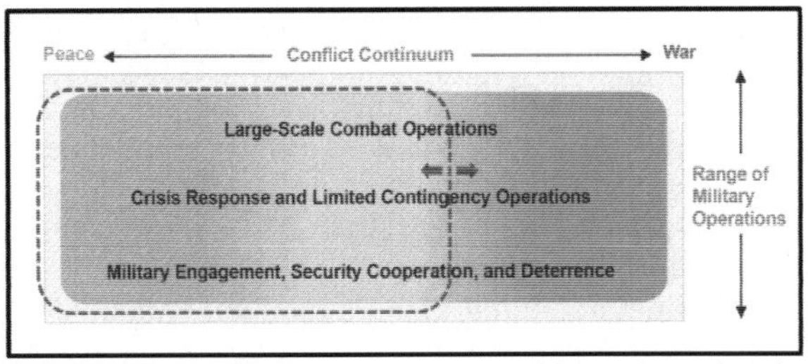

Figure 1: The Multi-Role Support Ship and Notional Operations across the Conflict Continuum.[4]

Further, military doctrine has become more collaborative where many responses to a given scenario involve multinational coalitions. Joint and combined operations involving many nations, agencies and non-governmental organizations drive the need for command and control, headquarters staff amenities, and planning and coordination facilities. Civil-military cooperation is now much more formally addressed as a requirement of contemporary military planning and the facilities to support effective interagency cooperation are vital.

A balanced fleet mix that includes combat capable vessels, ships that provide on-water support to para-military and constabulary functions, as well as other assets capable of delivering the soft-power elements needed in today's operational reality, is important. This includes the capacity to sustain a presence in an operating area, bringing substantial supplies and equipment into theatre, providing a self-contained afloat headquarters facility and getting boots on the ground to manage mass casualty situations, and provide security and safety to local populations.

Today, as we see the re-emergence of great power rivalry and a resurgence of traditional war-fighting capabilities in current force structures, the situations that have given rise to navies exploring the capability-mix inherent in the MRSS are not going away. This is the case both for the larger navies seeking to maintain a balanced force structure and for smaller navies equipping themselves to deal with regional situations and/or to contribute globally through specialization in niche capabilities. All navies need to consider how cost-effective solutions can be provided by some form of MRSS.

4. THE RATIONALE FOR THE MRSS

While the reasons for acquiring a multi-role asset are numerous, it is suggested that the rationale for any government will typically involve some combination of the following three key factors: operational flexibility, doctrinal relevance and affordability.

4.1 OPERATIONAL FLEXIBILITY

This refers to having flexibility within an operating cycle (between major dockings and depot-level maintenance periods) to respond to an urgent or emergent operational tasking. That is, having the ability to rapidly configure the vessel and mobilize for any given operation with the required mission payload, cargoes and specialist personnel embarked. When combined with an ongoing training and exercise program, this gives the government options to deploy quickly in response to any situation and allows for a high-readiness and agile posture on the part of the vessel operator. The inherent attributes of the MRSS that lend themselves to providing this operational flexibility include:

4.1(a) Ample Space and Deadweight Allowances

The MRSS is designed with large open and reconfigurable multi-use compartments and a sizeable deadweight capacity for cargo and mission payloads that lends itself to maintaining a flexible posture. The density for an embarked mission package will not vary radically whether transporting a mechanized infantry battalion or heavy equipment, vehicles and relief supplies for disaster response.

4.1(b) The Commercial Vessel Pedigree

The MRSS platform can be operated by a relatively small crew, often in a two-crew rotation system and, in many organizations, a civilian or hybrid civilian/military complement. Facilities for routine mission personnel, embarked as needed, to support capabilities such as helicopter operations and special forces, are also provided. Additionally, surge capability to provide hotel and support services for larger numbers, albeit in more austere accommodations, are also typically included.

Being based upon commercial ship designs, and being crewed and arranged in this manner, means that the operational availability that can be achieved is very high when compared with the maintenance, training and exercise time required to force generate a high-readiness complex military weapons platform.

4.1(c) Autonomy, Range and Endurance

These ships are designed with the endurance and range necessary to transport cargoes to remote and distant locations. The MRSS needs to operate autonomously in varied environments for extended periods. These vessels can also be pre-positioned within regions that are prone to natural disasters while undertaking other functions, such as training, exercise or patrol duties, and remain on station for extended periods of time. Specific mission payload items and specialist personnel can be mobilized and flown-in to a port facility and embarked en route to a specific incident to augment a standard contingency load-out when pre-positioned or forward deployed.

4.1(d) Mission Payloads

The MRSS can be designed and built with space, weight and power reservations specifically set-aside for the routine embarkation of modular mission payloads and/or force packages. A common interface specification (e.g. deck loading, securing arrangements, couplings and connections, power, data, securing arrangements) is provided to ensure maximum mission flexibility. If the payload can conform to the interface specification, then it can be embarked and operated from the ship. Mission bays, working decks, reconfigurable multi-use compartments, modular or containerized units and general-purpose spaces are included in this concept of a wholistic mission payload system. The main feature is that only minor preparation is necessary, permitting the rapid embarkation of mission payloads.

These modular items can be maintained ashore as part of the ships equipment list but only installed and set-to-work when specifically needed for a particular mission. In this way space and weight is not allocated to equipment that may only be needed occasionally.

4.2 DOCTRINAL RELEVANCE

A challenge always exists in delivering a contemporary operational capability that is timely, relevant and enduring. Arcane and lengthy government procurement processes, when set against the back-drop of the constantly evolving operational context described above, create a tension that is often very difficult to reconcile. This contributes to requirement creep that seems to impact many defence procurements, creating delays and cost overruns as new and emergent requirements are added during the latter stages of the acquisition process. Since procurement systems are traditionally slow to adapt and cannot keep pace with the prevailing operational context and technological advances, then one possible solution would be to ensure that what you procure transcends the operational doctrine of the moment. Navies need vessels that are resilient to doctrinal shifts over time and are future-proofed by design. The attributes of the MRSS that ensure continued operational relevance over the life of the asset include:

4.2(a) Scalable and Reconfigurable

What is embarked to perform a specific operation is scalable to the needs of that operation. Only the mission payload and specialist personnel need be embarked to complement the MRSS' core personnel and embedded capabilities. The MRSS can support missions ranging from lengthy patrol conducting anti-piracy operations or sanctions enforcement, to tactical sealift and amphibious landings, within a regional theatre of operations. In the first example, it may act as a mothership for maritime interdiction operations using small craft and helicopters, while in the second it would embark a mechanized infantry unit.

The MRSS is readily reconfigurable using moveable bulkheads, tracks, panels and hoistable decks. These features, often referred-to as flexible infrastructure, are easily integrated into this type of ship given their size and typical arrangement. When combined with modular furniture and fixtures, large spaces on the ship can be sub-divided for purposes such as refugee processing, triage for a mass casualty situation, stowage of segregated cargoes or used as surge accommodation for embarked personnel.

4.2(b) Adaptable

The RAND Corporation supports the argument that geopolitical uncertainty strengthens the case for adaptable ship design making it desirable to ensure adequate margins exist for ship services, namely power, cooling, personnel amenities, space, and bandwidth.[5] The MRSS design approach provides for these margins. For amphibious capability, they need to have the facilities to accommodate large cargoes and be able to embark many supernumerary or transient personnel. To provide command and control functions for an embarked incident command or forward deployed headquarters staff, the required bandwidth needs to be built-into the ship to accommodate administrative and tactical computer networks, communications systems, and data fusion needs.

The MRSS is also adaptable by re-allocating space, such as unoccupied lane-metres on the vehicle deck or space on the upper decks, to other temporary purposes. For example, additional power can be added through containerized generators and energy storage modules, potable water production capacity augmented with demountable reverse-osmosis desalination units plumbed to a fitted sea-bay, or skid-mounted air conditioning plants connected to auxiliary connections fitted on the chilled-water system.

4.2(c) Upgradeable

The notion of upgradeability is closely linked to adaptability but intends to specifically address the insertion of new technologies and/or the replacement of existing systems or equipment with the latest updates. To accomplish this the ship design employs open and flexible systems architectures, along with the integration of commercial, off-the-shelf equipment that utilize standardized interfaces. Attention to maintenance envelopes and removal routes during the design process will be important to maximize this potential.

The concept of 'fitted for but not with' could also apply here in that the ship could be built in anticipation of potential future missions and the downstream installation of equipment such as specialised lifting appliances, additional generators or water-making facilities.

4.3(d) Common Hull with Optional Topside Blocks

For larger navies, or potentially multi-national programs, there is the potential to consider the MRSS as a family of ships, using a common platform with a range of topside configurations. Designing and building the ship with this block philosophy also facilitates mid-life modification if required to address changes in doctrine, missions and technology.

As such, the MRSS could be retrofitted and repurposed between different primary roles. If outfit density, stack-up lengths, HVAC and electrical loads are generally consistent then the business case to undertake the work and repurpose the vessel could be examined. Presuming of course the condition of the vessel and remaining useful life of the underlying hull and main machinery provide for an adequate timeframe over which to realize a return on the investment.

Figure 2: The retrofit of topside blocks for repurposing between primary roles

If this possibility is contemplated at the outset of the program, then the ship can be designed and built in blocks that facilitate this type of conversion in the future.4.3 AFFORDABILITY

The best value solution is one which provides the required level of mission effectiveness for the lowest total ownership cost over the entire lifecycle of the asset. The three key themes in addressing the best-value question are:

4.3(a) The Fleet Mix – Doing More with Less

In the first instance, affordability should be assessed in terms of the overall fleet mix. Is it better to own a specialized hospital ship, a fleet replenishment vessel and a dedicated sea-lift capability or three MRSS that can cater to all those missions, albeit perhaps not to the same extent as a specialized vessel? Given commercial crewing regimes and utilization rates, could the capability resident in three specialized vessels be accomplished with just two MRSS?

Acquiring multiple MRSS also lends a measure of redundancy to a fleet operations plan knowing you have a like-asset able to backfill another in the event of unplanned downtime.

4.3(b) Reduced Capital Expenditures

Over the last four decades the inflation rates associated with naval shipbuilding has consistently exceeded twice that of normal inflation indices.[6] An MRSS designed, built and operated to commercial standards and maintained in Class will be inherently less expensive to acquire and operate than a traditional military asset.

Given the commercial pedigree of this ship type, using state-of-the-market, type-approved (i.e. non-developmental) equipment, that is commercially available off the shelf, will streamline integration, installation and set-to-work. It will also come at a lower initial price point and with a well-established support program, resulting in lower up-front integrated logistics support costs.

4.3(c) Reduced Operating Expenditures

Higher utilization rates and operational availability may allow the same number of operational sea days to be achieved with fewer assets. Additionally, if a commercial or hybrid crewing and operating model is employed, and, given the potential for achieving higher utilization rates, the cost of ownership per operational day is driven even lower. Further, owning multiple identical assets, or cooperating with other like-minded nations in a parent-navy arrangement, will allow for class management and economies of scale to be realised for training, sparing, and maintenance.

Of course, assessing the mission effectiveness of all options – that is, the how, where and when of bringing operational capabilities to bear – is an important part of the calculus in determining best value.

5. THE PITFALLS AND ACHIEVING AN OPTIMAL OUTCOME

5.1 AVOID INCOMPATIBLE MISSION SETS

The main situation to be cognisant of is ensuring that the vessel, as designed, is competent to perform the roles it is required to undertake. That is, avoiding the "jack of all trades and master of none" outcome where multi-purpose is synonymous with multi-useless. Fundamentally different types of ship have different design drivers, and this leads to great differences in configuration in the underwater hull forms, above water hull and superstructure, propulsion plant and auxiliaries, and hotel services for the persons on board. In some cases, they are also subject to differences in regulatory requirements, which can also affect design outcomes significantly.

The objective is to identify and avoid trying to combine fundamentally incompatible mission sets.

5.2 FINDING THE RIGHT BALANCE

Any multi-mission ship is a compromise between design drivers that will inevitably conflict to some extent. It will be sub-optimal in every mission, to a degree reflecting the priority given to each capability in the design. Despite this, many naval vessels today are considered multi-mission, and the disadvantages are accepted as being smaller than the costs of trying to justify a necessarily larger fleet of more specialised ships. Aiming for the gold standard in every mission requirement will drive size and cost. Walking back on current capabilities in any area will need careful consideration and a thorough analysis of operational effectiveness.

The hull must provide enough buoyancy to support the weight of the ship itself and the weight of everything it is required to carry. Bulk cargoes are dense, people are not, but containers and vehicles fall in between. Accordingly, the hull form design must provide enough buoyancy for the maximum displacement while still meeting speed and sea-kindliness requirements to be operationally effective.

Another challenge will be to arrange ballasting and subdivision to address a range of deadweight and stability conditions in different mission configurations. Considerations such as damage tolerance and vehicle loading lead to people and vehicles being carried high in the ship, increasing the stability management challenge.

Depending on the hull form, ballast and stabilization systems may be needed and pose other challenges for the hull designer. If the deadweight requirements, along with outfit and load-out densities, are relatively consistent then an optimum hull form and propulsion arrangement can be developed. Powering needs to consider a range of missions and may use hybrid solutions for maximum flexibility and capability.

Internal layouts need to consider work flows for safety and effectiveness. How the ship is arranged internally can be optimised around both the higher priority missions and an analysis of the work flows required to safely execute those missions with the desired level of effectiveness.

5.3 CREW COMPETENCY PROFILES

One of the biggest drivers of through-life costs is personnel. In an era, particularly with commercial ship operations, where an emphasis has been placed on automation and autonomy with a view towards reducing crew sizes, building a ship that has many varied and specialised functions will require many crew competencies. This would tend to drive crew sizes higher and will create a training burden that could challenge normal occupational structures and crewing models.

Maintaining a required complement of suitably trained and experienced personnel is a vital component of both readiness and mission effectiveness and cannot be divorced, or considered separately from, the asset itself. Accordingly, the crew, and the enabling competency profiles, should be considered as an integral part of the capability when requirements are initially set and then more fully considered during early stage concept design work. The competency profiles will change and evolve as those requirements are more fully defined and validated. Part of the solutions to resolving any challenges of this nature could be to consider specialist personnel as being an integral part of the mission payload or modularity concept. That is, when the mission payload is embarked the specialist personnel will also embark and augment the core crew complement.

The bottom line is that a work flow and crewing analysis should be performed early and then refreshed with each major design review milestone.

5.4 EARLY DESIGN COLLABORATION

When developing a multi-role vessel design the best contracting strategies are those that promote early-stage design collaboration. Integrated early stage collaboration between expert ship designers and expert ship operators is vitally important to ensuring realistic expectations are established and an optimum outcome is reached. In the case of HADR operations, adding expert end-users, such as non-governmental organization and aid agencies, into the discussion is also important. Those who have experienced the circumstances in which the vessel will be operated are key to developing an optimally arranged and workable solution.

Designing a multi-mission platform will be challenging and will inevitably require trade-offs. Choices will need to be made and this needs to be done within the context of mission effectiveness and a clear articulation of operational priorities. To accomplish this in a process that involves many stakeholders, likely with competing views and motives, and a long list of multi-mission requirements, the designer has two key tools in the toolbox – set-based design and scenario-driven design techniques.

5.4(a) Set-based Design

A set-based design approach, where modern ship-design tools and methodologies are applied concurrently, is ideal for the MRSS. Having the ability to leverage a family of parent vessel concepts will provide the experienced designer a viable basis around which to bound the design-space and establish a starting design-set. As opposed to point-design, this design-set allows for multiple options to be considered for as long as possible during the concept development process. These design options are simultaneously explored with sub-optimal choices eliminated progressively over time as clarity is gained and evidence-based decisions are made.

This in turn will permit a dynamic work environment to quickly and effectively respond to changing requirements as discussions with end-users ensue and their own unique combination of needs are described in detail. It enables greater flexibility and a more agile design process by committing to technical solutions only after requirements are reconciled and validated based on mission effectiveness and cost.

5.4(b) Scenario Driven Design

Scenario driven design involves the application of operational scenarios developed by ship operators and end-users to help the designer understand operational intent. This will help guide the design team towards a purpose-driven whole-ship and system-level design based upon operational imperatives, procedures and workflows. Scenarios need to be realistic and relevant to afford multiple perspectives of workflows and provide a tangible and effective way to engage end-users in early design stage discussions. This collaborative approach will allow the designer to assimilate and synthesize a great deal of information quickly by providing real time context for operations that is in turn tempered by realistic expectations regarding technical feasibility and affordability. Scenario driven design will help ensure alignment between operational intent and design intent before detailed design and engineering work commences.

6. CASE STUDY: DEPLOYING THE MRSS ON HADR OPERATIONS

6.1 THE MULTI-ROLE SUPPORT SHIP

The VARD 7-313 is a multi-purpose vessel designed as a flexible platform to aid in force projection, maritime special operations, EEZ patrol, and humanitarian assistance. Typical for vessels of this type, the vessel has significant capabilities for offloading heavy equipment, carrying cargo, transporting troops, launching boats, and supporting aviation operations.

Figure 3: The VARD 7-313 HADR Variant

The internal roll-on/roll-off facilities accommodate a variety of equipment including tanks, trucks, and ISO containers which can be efficiently loaded from a pier via ramps on the stern and vessel's side. Two full-breadth cargo holds located beneath the vehicle deck provide extra storage capacity, while protected areas forward house expansive troop accommodations providing safe and comfortable passage for embarked personnel. Large open deck areas on each side of the vessel provide secure stowage for two 15 m landing craft and two 11 m rigid-hulled inflatable boats. Aviation facilities include a flight deck capable of landing two medium lift helicopters and a hangar accommodating a total of four.

MAIN PARTICULARS		
Length overall	130.0 m	436'-4"
Length waterline	123.9 m	406'-6"
Breadth moulded	24.0 m	78'-9"
Depth main deck	9.0 m	29'-6"
Design draft	5.4 m	17'-8"
PERFORMANCE		
Speed	20.0 kn max	
Range	8,000 NM @ 16 kn	
Endurance	30 days	
CLASS		
BUREAU VERITAS I✠HULL MACH RoRo Cargo Ship, AUT-UMS, AUT-CCS, AUT-IMS		

Figure 4: VARD 7-313 Principal Particulars

Expansive medical facilities are readily accessible from the flight and boat decks. The substantial internal area is dedicated to configurable offices and operations areas to support various mission types. As shown in Figure 5 the VARD 7-313 offers a high degree of utility with a substantial allocation to of area and volume to cargo and both fixed and flexible mission spaces. There are 475 lane-meters or 1,330m^2 of total modular area on the vehicle deck plus another 740m^2 of reconfigurable flexible mission space.

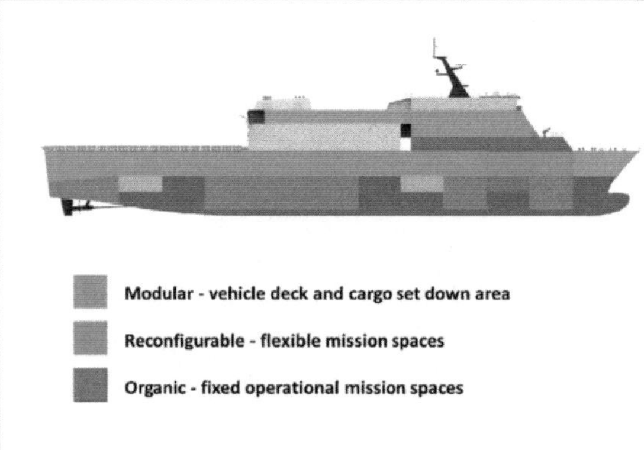

Figure 5: VARD 7-313 mission space distribution

6.2 HADR CONCEPT OF OPERATIONS

There are several different references that describe a framework for how to plan and execute a humanitarian assistance operation. Whether one examines the UK Ministry of Defence Joint Doctrine Publication 3-52[7], the ASEAN Forum Strategic Guidance for Humanitarian Assistance and Disaster Relief[8], or any number of other documents, the approaches are all very similar. For the purposes of this case study, the execution of a typical HADR operation will consist of the following five phases: Mobilization; Immediate Life Saving; Needs Assessment; Stabilization; and, Recovery.

6.3 OPERATIONAL SCENARIO

The generic scenario describes a category 5 hurricane that makes landfall on a small island nation in the Caribbean Sea with devastating results. Coastal towns and infrastructure are destroyed in high winds and tidal surges, including port and harbour facilities, and local airports and airfields have sustained heavy damage. Infrastructure, including the power grid, communications networks, and municipal water and sewer facilities have sustained significant damage and are inoperable. Many local roads are impassable. While information is sketchy and difficult to verify, media is reporting widespread loss of life, significant injuries as well accounts of many people whose whereabouts are unknown.

The vessel was pre-positioned and forward deployed conducting training exercises off the Atlantic seaboard of the United States. The ship is crewed by a core complement and is outfitted with a standard HADR contingency load-out. This includes medical and relief supplies, 5-ton transport trucks, water and fuel bowsers, heavy equipment, lighter reconnaissance vehicles and ambulances, a modular field hospital, deployable shelter systems, a mobile command post and two demountable reverse osmosis desalination plants. One light reconnaissance helicopter and two heavy-lift helicopters are embarked along with flight crews and aircraft maintenance personnel. A deployable modular raft ship-shore connector system is also embarked.

Authorities have been tracking the developing weather situation and a decision was taken early by national authorities to task the ship to prepare for a major hurricane response. Models are predicting the storm will make landfall as a category 4 or 5 hurricane within the next 48-72 hours.

6.3(a) Mobilization

Immediately upon being tasked the ship steamed directly to a Southern US port to embark additional equipment and specialist personnel flown in to meet the ship. A troop of military combat engineers, a military policy security detachment, urban search and rescue specialists and additional medical surgical and trauma teams joined the ship. Additional medical and relief supplies were also embarked along with additional temporary shelter systems and construction materials to aid in the repair of critical infrastructure. Personnel from various non-governmental organizations and aid agencies were also invited to join the ship for the voyage south. Headquarters and military planning staffs joined the ship along with several liaison officers trained in civil-military relations. A third heavy lift helicopter from an allied navy was also assigned to the ship.

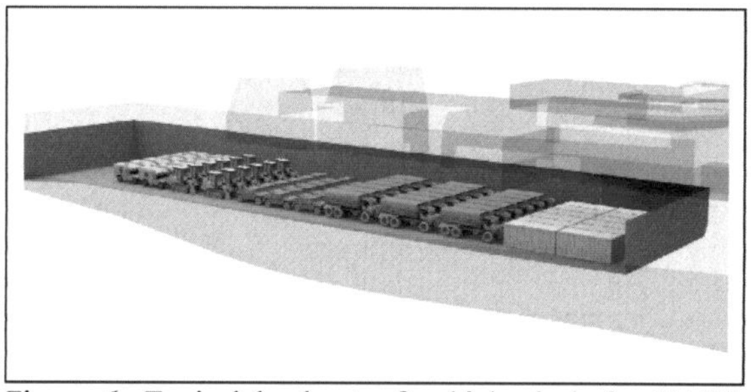

Figure 6: Typical load-out of vehicle deck for HADR operations

Once loaded, a route, avoiding the worst of any residual weather was planned, and the vessel sailed with all dispatch to the affected area.

6.3(b) Immediate Life Saving

En route to the area, the ship is tasked as a vessel of opportunity by local SAR authorities to respond to any cases resulting from the storm activity in the area.

Upon arrival on scene the focus in the first 48 hours is life saving and harm minimization. Off-board systems are immediately deployed: helicopters are dispatched to conduct aerial reconnaissance and the ship's boats are deployed to assess the situation in the ports and along the coast. Liaison officers are transported ashore to establish contact with the local emergency management authorities. Helicopters and the ships boats are tasked with search and rescue operations.

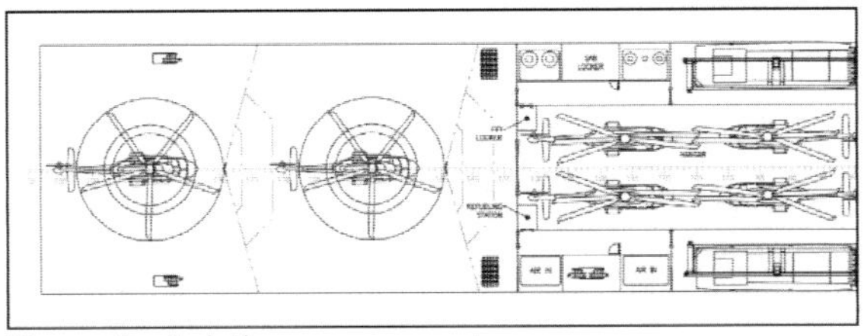

Figure 7: Aviation arrangements

The arrival of the ship brings a self-contained fully staffed and operational NATO Level II hospital facility to the scene for casualty receiving and treatment. Medivac is conducted as needed to transport urgent cases to the ship where they are received in triage and treated appropriately. Those requiring emergency surgery are treated and then cared-for in the ICU ward. Flexible mission space is configured as a recovery ward and to accommodate those patients not requiring intensive care.

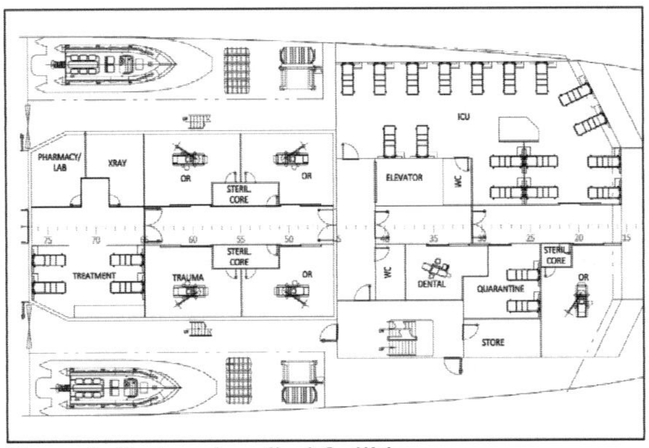

Figure 8: On-board medical facilities

Security detachments can be deployed to work in cooperation with local authorities to help maintain order in what has likely become a very chaotic situation.

6.3(c) Needs Assessment

The on-scene commander, having now established direct liaison with local authorities, and the planning staff have begun to assess the information they have received in the onboard command and control facilities. NGO and aid agency staff are working from assigned on-board office and planning spaces in a coordinated and fully integrated manner. A priority list of urgent needs is established, plans created, and resources are allocated.

A decision is made to create a suitable beach head for landing equipment and marshalling relief supplies ashore. Military engineers are deployed to scout a suitable landing beach and prepare it for landing operations. Heavy equipment and supplies are landed using the ships landing craft while the crew begins assembly of the larger modular raft ship-shore connector in preparation for larger scale landing operations.

A security perimeter, secondary command post, marshalling area and medical receiving facility are established immediately inland from the beach head. Critical aid supplies are prioritized and begin getting transferred ashore for distribution from the marshalling area.

6.3(d) Stabilization

Over the course of the first 5 days, having addressed the most urgent needs, stabilization efforts begin to preserve life. Aid supplies continue to get offloaded and distributed as needed. Temporary shelters are disembarked and set up to provide safe and secure accommodations for displaced persons. Medical cases are assessed ashore with urgent cases referred to the onboard medical facilities for treatment.

One of the most critical elements will be the provision of potable water. The ship can provide potable water directly from the onboard water storage tanks to bowser trucks and trailers on the vehicle deck which are then transported ashore and distributed as needed. To augment the ships production capacity demountable reverse osmosis desalination unit can be plumbed to a manifold connected to a sea-bay. These demountable units will then be transitioned ashore to support water purification needs locally where needed.

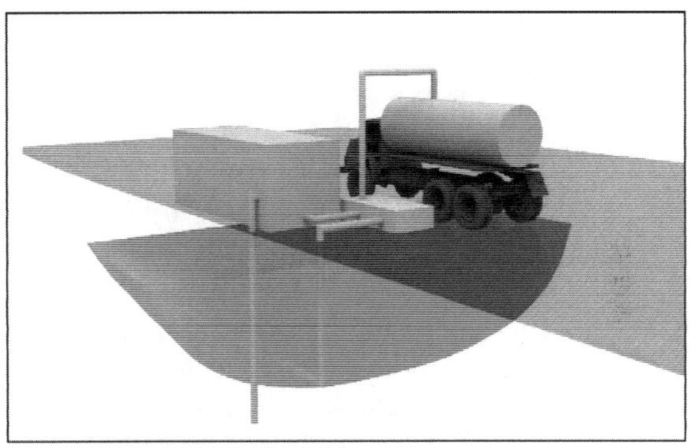

Figure 9: Provision of potable water

6.3(e) Recovery

Over the first 4 weeks the engineering teams, their equipment and building materials are utilized in the remediation of sea port and airport facilities. This will be performed on a priority basis to allow airlift and sealift operations to begin in earnest. Reconstruction of local medical facilities and restoration of utilities will ensue, and ambulatory medical cases can begin to be transferred ashore as local infrastructure and services are restored.

Once infrastructure has been restored and command post operations fully transitioned ashore the ship can then undertake resupply missions as needed. Transition of the operation to local authorities is undertaken and will occur as rehabilitation continues.

7. CONCLUSION

The best answer to any question is often the simplest. In that vein, perhaps the emergence of the contemporary MRSS is simply a response to the prevailing operational context within which many of today's naval forces around the world operate, where necessity becomes the parent of invention. While that might seem like a glaringly obvious statement the proliferation of the MRSS as a concept in the minds of naval planners, on the drawing boards at design houses, under construction in shipyards and alongside at naval bases around the world cannot be a coincidence.

The MRSS vessel type is particularly adaptable to ever-changing prevailing operational needs of small and large navies alike. The vessel will provide an ideal platform from which to mount a comprehensive and self-contained response to HADR operations. An optimal solution will produce the benefits of operational flexibility, doctrinal relevance and affordability. These benefits can all be realized if the procurement approach permits early design stage collaboration between expert ship operators and expert designers.

REFERENCES

1. Naval Technology, 'Type 26 Global Combat Ship (GCS) Programme', *Verdict Media Limited*, https://www.naval-technology.com/projects/ global-combat-ship-gcs-programme/ (accessed 14 May 2019).

2. Federal Fleet Services, 'Resolve Class AOR Fact Sheet', http://federalfleet.ca/wp-content/uploads/2016/07/Resolve-Pamphlet.pdf (accessed 14 May 2019).

3. Katharine Hayhoe et al., '20014 National Climate Assessment', *U.S. Global Change Research Program*, https://nca2014.globalchange.gov/highlights/overview/overview (accessed 14 May 2019).

4. Joint Publication (JP) 3-0, 'Joint Operations', *United States Joint Chiefs of Staff*, 2017.

5. John F. Schank et al., 'Designing Adaptable Ships', *RAND Corporation*, 2016.

6. Mark V. Arena et al., "A Macroscopic Examination of the Trends in U.S. Naval Ship Costs Over the Past Several Decades', *RAND Corporation*, 2006.

7. Joint Doctrine Publication (JDP) 3-52, 'Disaster Relief Operations Overseas: the Military Contribution', *UK Ministry of Defence*, 2016.

8. ASEAN, 'Strategic Guidance for Humanitarian Assistance and Disaster Relief', *ASEAN Regional Forum*, 2010.

AUTHORS BIOGRAPHY

Derek Buxton holds the current position of Director Government Programs at Vard Marine Inc. He is responsible for global business development in the government vessels market segments including offshore patrol vessels, icebreakers, military survey ships and naval auxiliary vessels. His previous experience includes a combined 32 years of service as a Marine Engineering Officer and Senior Procurement Executive with the Royal Canadian Navy and Canadian Coast Guard. In 1992 he deployed with HMCS Preserver as the Deputy Engineering Officer on Operation Deliverance in support of humanitarian efforts in Somalia.

DEVELOPMENT OF A TWIN STERN RAMP LAUNCH & RECOVERY SYSTEM FOR RIGID HULL BOATS ON A MULTI-ROLE NAVAL VESSEL

N L Koh, D B Wong, Y M J Tan, B H Chew, T S Gan, Defence Science and Technology Agency (DSTA), Singapore

SUMMARY

Deployment of fast response craft at sea is essential for a range of littoral missions including maritime security and search-and-rescue. Traditional launch and recovery of fast response craft like rigid-hull boats (RHB) employ side davits and cranes. In light of evolving technology, multi-role navy ships like the Republic of Singapore Navy's *Independence*-class Littoral Mission Vessel have incorporated stern ramp Launch and Recovery Systems (LARS) to overcome the slower deployment times associated with cranes and davits. Twin stern ramp LARS are suitable for requirements of multi-RHB operability, while achieving faster craft launch and recovery with fewer deck crew required. The paper identifies the design considerations, safety considerations, and qualification process for twin stern ramp LARS.

NOMENCLATURE

m	metres
deg	degree

1. INTRODUCTION

The deployment of fast response crafts such as Rigid Hull Boats (RHBs) off many modern navy and coast guard ships are essential to the success of seagoing missions. RHBs are reconfigurable and offer a cost-effective alternative to conduct missions ranging from maritime enforcement, salvage, drug- and migrant interdiction, and rescue insertion/extraction. These missions entail the safe transfer of military or civilian personnel between small boats and the mothership.

The configurable nature of RHBs combined with superior speeds – typically at least 40 to 50 knots – make for an attractive mission asset. Quick deployment and recovery of the embarked RHB is essential for shorter mission response and turnaround time. Traditional over-the-side launch and recovery systems (LARS) typically require 15 minutes to deploy or recover a RHB. Deck crews from the mothership are required to handle painter lines attached to the RHB to prevent uncontrolled movement throughout the launch and recovery process. At higher sea-states, the mothership has to create a lagoon of flat waters, by performing an inward turning-circle manoeuvre, to aid the launching. In a rapid-response situation such as drug interdiction, those 15 minutes make all the difference between successful interdiction and futile pursuit.

Navy and Coast Guard vessels seeking shorter RHB deployment time or multi-craft operability are turning to stern ramp LARS to meet their operational requirements. Advantages of stern ramp LARS are shorter deployment times, and lower manpower costs associated with launch and recovery.

Launching a RHB using a stern ramp involves opening the stern gate and releasing the RHB. The RHB either slides down the ramp under its own weight or via hydraulic/pneumatic rollers to control its launch speed. Once clear of the mothership, the RHB drives away under its own power. During recovery, the RHB is either driven or winched aboard the mothership's ramp and quickly secured.

Research on more effective means to aid RHB recovery could also be extended to applications for unmanned surface vehicle (USV) and unmanned underwater vehicle (UUV) recovery. Cosmo [1] proposed a brightly-coloured towline to improve visibility and design of a compliant member to reduce loads and risk of latch damage. Petersen [2] explored a probe-receiver device for towing. Galway [3] outlined design considerations and design loads of a towline capture latch systems. Kilbourne [4] reviewed multiple latch designs to snag a towline trailing from the mothership.

LARS elements to optimise include the size and angle of ramps and doors, door design, fenders, water management, and interface between the mothership and the RHB. Dallinga [5] concluded that scale model tests were useful for validation of LARS system safety, water management and optimal heading. Carette [6] conducted time domain simulations in addition to model tests to investigate optimal launch and recovery design. McTaggart [7] used simulation models to investigate the effects of hydrodynamic interactions between a small boat and mothership. Schmittner and Carette [8] discussed the merits of model tank tests in modelling the mission-representative hydrodynamic environment for launch and recovery operations.

On stern ramp design enhancements, Morriseau [9] proposed an additional stern notch with garage door on a patrol boat, with rub rails to reduce damage on rapid response craft. Mau et al [10] proposed submerged ramp extensions with goalpost-like floats to aid recovery guidance. Their investigation also concluded that cut outs on the ramp extension reduces stern flow disturbance.

This paper documents the findings of different launch and recovery systems for fast response boat and discussion on the design and development from model tests carried in MARIN, Netherlands, both Harbour Acceptance Test (HAT) & Sea Acceptance Test (SAT) in Singapore. It also provides prospective future directions on the development of the systems.

2. DESIGN AND DEVELOPMENT

2.1 Several LARS designs were evaluated against the requirements for the efficient launch and recovery of multiple craft types from the mothership. It was also desired for the least number of deck crew to perform launch and recovery operations.

2.1.2 DAVITS- AND CRANE-TYPE SYSTEMS

Davits and crane-type LARS were assessed to have a lower impact on ship design and a smaller space requirement compared to stern ramps as illustrated in Figure 1. The relative disadvantage of davits and crane-type LARS were lack of operability at higher sea-states and larger deck crew numbers required to assist with launch and recovery operations.

Figure 1: A boat davit on the Izumo-class helicopter carrier

The types of davits considered were:

a. Crescent Davits. A two-armed davit that is moved outboard of the ship by a hand crank or an air-powered motor. Once outboard, the RHB is lowered by lifting the brake handle and allowing the weight of the boat to lower itself.

b. Pivoted Gravity Davit. A two-armed gravity davit that works in two distinct stages. After the RHBs is ready to lower, the brake is released and the winch pays out. The davit arms pivot into the outboard position and come to rest against mechanical stops. The RHB is lowered onto the water during the second stage.

c. Trackway Gravity Davit. A two-armed gravity davit that works similar to the *pivoted davit* except the arms slide down an inclined trackway on rollers.

d. Single-Arm Trackway Gravity Davit. A one-armed version of the *trackway gravity davit*. The single arm is mounted in the centre of the davit area and designed for a motor whaleboat equipped with a bail that bolts into the lifting pad-eyes in the bilge. The use of the bail and quick-trip hook allows for faster, safer launch and recovery operations at higher sea-states than other davits.

e. Slewing Arm Davit. A slewing arm is rotated around a turning centreline and comes with lowering functionality customised to individual demands. A cradle fitted onto the davit aids in recovery.

2.1.3 STERN RAMP LAUNCH AND RECOVERY SYSTEM

Stern ramp LARS allowed for RHB operations at higher sea-states than its davit and crane counterparts, at the expense of larger footprint and design impact to the mothership. The team assessed that stern ramp launch systems were inherently safer, taking into account associated deck crew hazards, seamanship, and stability of the RHB throughout launch and recovery. On robustness, the stern ramp LARS could perform a one-off RHB launch without stored power in the event of black-out, through gravity-assisted descent. Launch and recovery operations for stern LARS required fewer deck crew than other designs. The following stern ramp launch designs were considered:

a. Tilt frame with sliding bed. A frame is tilted up by hydraulic arms before extending the sliding bed into the water. A recovery line is attached to the RHB once it is driven into the sliding bed.

b. Capture hook. A line is released from the mothership and attached onto the RHB. With this connection, a rigid capture head can be attached to the RHB for winch in operation. Alternatively, the RHB is captured by a fixed hook on a traverse system to pull the RHB up the ramp.

c. Extended ramp. An extended ramp is availed during launch and recovery to increase sill depth, which improved RHB capture by LARS infrastructure. Rollers on ramps are hydraulically operated to move the craft up/down the ramp with no need for recovery lines.

2.2 CONSIDERATIONS

The two main class of LARS were evaluated, and stern ramp launch systems were preferred based on the following considerations:

a. Safety. Davit solutions required the RHB to be suspended in mid-air with crew on board. This increases the risk of crew injury and equipment damage, especially at higher sea-states. The stern ramp solutions afforded safer operations as the movement of the RHB is constrained and not located at height.

b. Manpower. Davit solutions required more manpower for each operation due to the need for line handlers to control the movement of the suspended RHB.

c. Time for deployment. Stern launch and recovery was found to be quicker than davit and crane systems. The stern extension or doors may provide a funnelling effect to improve RHB guidance during recovery.

d. Simplicity of operation. The stern launch solutions provide a simpler method of operations whereby RHB can be launched by gravity or in a controlled manner compared to managing swinging loads during operation.

e. Redundancy. Stern ramp solutions can be fitted with backup winches to provide another means of recovery of the RHB in the event of failure of the main hydraulic rollers.

2.3 SELECTION AND DEVELOPMENT OF A STERN LAUNCH AND RECOVERY SOLUTION

Most naval vessels achieve multi-craft launch ability via cranes and well docks. The team pursued an alternative approach to simplify multi-craft launch and recovery without the increase in deployment time and deck crew associated with cranes and well docks. A series of technical assessments, cognitive walkthroughs and crew interviews identified the *twin stern ramp* solution as the most suitable system for the multi-role naval vessel's unique operating requirements. The design and development journey unearthed latent advantages of the well-integrated twin stern ramp shown in Figure 2:

a. Flexible space utility. When either or both stern ramps are not in use, the deck space can be utilized to bring in more capabilities.

b. Weight. The overall weight for stern ramp is lower than the tilt frame solution, therefore unlocking more payload capacity for the mothership.

c. Flexibility. The rollers for stern ramp solution can be adjusted to conform to any hull form due to its 3-axis rotation to accommodate the beam of the RHB, depth of keel, and angle of hull form.

d. Deployment time. The design of the stern door serves to extend the ramp and increases sill depth for recovery. The mothership is not required to take in ballast to meet the loading requirements for recovery operation.

Figure 2: Stern ramp LARS for a multi-role vessel (Starboard ramp shown)

3. MODEL TESTS

Model tests afford the opportunity to cost-effectively test systemic variations of designs. The objective of the model tests are to indicate the operational envelope and limitations of the RHB, mothership and LARS in the envisaged operating conditions. Limitations of davit-type LARS are assessed by evaluating the accelerations at the operator location and relative motion between the mothership and waves. This data is traditionally evaluated with established seakeeping motion analysis tools. The numerical evaluation of stern ramp LARS was more challenging and with limitations, due to the complex interaction between the mothership's stern wake, propeller wash and incident waves [8].

3.1 MODEL TESTING FOR TWIN STERN RAMP LARS DEVELOPMENT ON A MULTI-ROLE VESSEL

The development of the twin stern ramp LARS involved extensive model tests in to investigate the operational performance of the RHB deployment system as illustrated in Figure 3. Test campaigns were conducted over two weeks to study the RHB deployment system. In total, more than 100 runs of launch and recovery operations in waves were conducted in the MARIN Seakeeping and Manoeuvring Basin (SMB). Model test data was collected and analysed to determine the indicative operational envelope and for RHB launch and recovery. Video recordings and photographs were also used to supplement quantitative data.

3.2 MOTHERSHIP AND RHB MODEL

The multi-role vessel (mothership) and RHB models used for the launch and recovery tests were manufactured by MARIN, with a geometric scale ratio of 1 to 12.173. All tests were performed with the same loading condition of mothership and RHB.

The mothership base model was also used for seakeeping assessments and powering tests for the multi-role ship. The mothership model featured a twin shaft design with two stock propellers. A pair of steerable rudders and long bilge keels were incorporated on the model. The mothership was equipped with a stern starboard RHB LARS. The deployment system comprised a ramp and an opened stern door. For all tests, the stern door was kept open by two hinges and a fixed side pipe. A force transducer on the fixed pipe measured forces borne by the hydraulic system. Two pairs of pods and four pairs of wheels were installed on the outer ramp door section to guide the model RHB, while eight pairs of wheels were installed on the ramp. All installed wheels were the free-rotating type.

The RHB model, as shown in Figure 4, was designed by MARIN based on dimensions on the envisaged rapid-response craft to be embarked on the multi-role vessel. The RHB was equipped with one water jet, which allowed a helmsman to remote-control its direction and speed.

Figure 3: Mothership model stern ramp LARS – stern quarter view

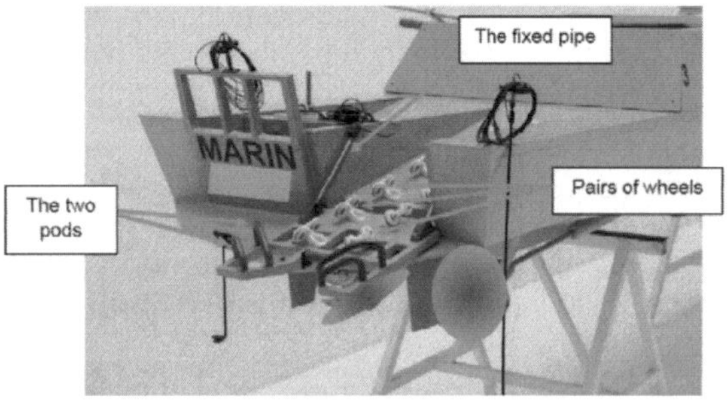

Figure 4: RHB model

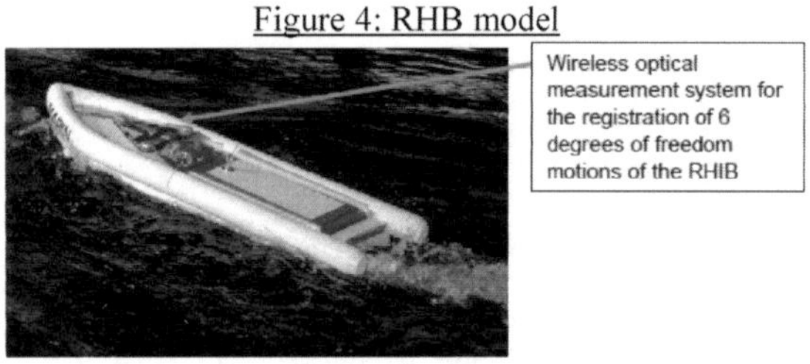

3.2 TEST ARRANGEMENT

The model tank tests were conducted in the SMB which measures 170m x 40m x 5m in length, width and water depth, respectively. A towing carriage travelled across the basin with the test models, carrying recording equipment and personnel.

Both models were self-propelled during all tests as shown in Figure 5. Connections between mothership and carriage comprised free-hanging electric wires for relay of measurement signals and power supply. The mothership model was manoeuvred via autopilot reacting on course deviation, rate of turn and lateral displacement. The RHB was completely wireless, and steered by a human helmsman who was positioned aft of the mothership on the carriage. The measured quantities comprised: basic six degrees of freedom motions of the mothership and RHB, wave elevation at the mothership's stern, mothership speed, and force borne by the hydraulic piston keeping the stern ramp door open.

Figure 5: Launch and recovery model tank tests

The relative motions of the RHB compared to the mothership were important to understand the operating envelope and outcome of hydrodynamic interactions during launch and recovery operations. Relative x-, y-, z-motions and acceleration between the mothership and RHB were derived from measured signals. Plotting relative x-motions against relative y-motions indicates the difficulty of aligning the RHB with the mothership. Plotting the relative z-motion against relative x-motion gave insight into the effect of hydrodynamic interactions at the stern of the mothership as illustrated in Figure 6.

Figure 6: A sample relative motion plot to investigate effects of alignment difficulty (left) and effects of hydrodynamic interaction at stern of mothership (right)

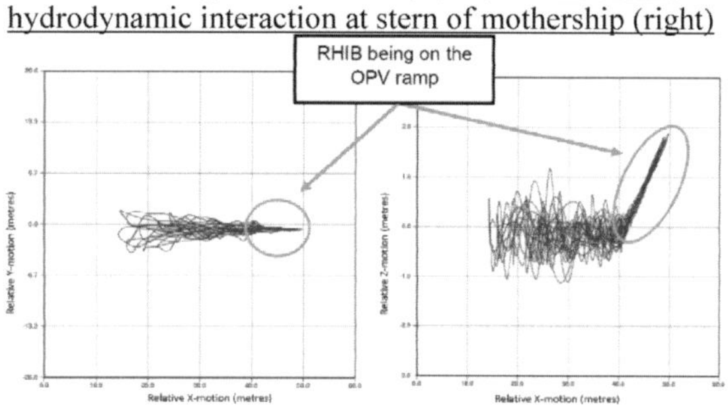

The sea-state definition and significant wave heights of the sea-states applied for this project was based on NATO STANAG 4194, consistent with model testing best practices of other navy vessels. A Pierson-Moskowitz spectrum was used to generate irregular wave conditions for mission-representative wave conditions up to sea-state 4.

The model scale of 1:12.17 used meant that time ran 3.49 times faster than in reality. The time for the human helmsman to react was therefore significantly shorter. This effect was compensated by positioning the helmsman aft of the RHB and mothership with a bird's eye view. In consequence, the model operational behaviour is assumed to be realistic. The quality of RHB control was dependent on helmsman experience and skill. In reality, the success of RHB launch and recovery operations were contingent on crew competency in parallel to the dynamic environment, operational conditions and safe system design. Test runs were conducted randomly in terms of the sea-states, wave height and directions, and ship speed. This was to ensure that the helmsmen does not become overly accustomed to the wave conditions which could skew the results.

Each test campaign was designed with sea-states varying from 2 to 4, in 6 different headings (0 deg, 90 deg, 135 deg, 180 deg, 270 deg and 315 deg), and different mothership speeds to replicate the real operations. The runs were repeated for Sea Sate 3 and 4, to obtain more results for comparison.

Safety of RHB launch and recovery was scored on a three-grade scale by members of the test team comprising representatives from the Defence Science and Technology Agency (DSTA), Republic of Singapore Navy (RSN), SAAB Kockums (then-ThyssenKrupp Marine Systems AB), ST Engineering Marine Limited (STEML), and MARIN.

4. ACCEPTANCE TESTS

The second phase of the LARS qualification process entailed full scale tests of the design with actual RHBs in order to validate the model tank test results. The second phase focussed on integration of the LARS-RHB assembly, potential in-theatre operating issues, and qualification of system safety mitigation measures. This effort was undertaken by Palfinger and STEML, with observers from the DSTA and the RSN.

4.1 LARS PROTOTYPE TESTS

The first of these validation tests began in September 2014, with STEML and Palfinger building a prototype LARS consisting of hydraulic roller pairs installed on a ramp simulating one of the multi-role ship's stern ramps, as shown in Figure 7. These hydraulic rollers were configured to launch and recover various small craft from the LARS prototype, including a 6m RHB. These tests were done to ascertain the viability of the launch and recovery of RHB in the full scale design before implementation on the ship.

Figure 7: Prototype LARS set-up

4.2 HARBOUR ACCEPTANCE TESTS

Harbour Acceptance Tests (HAT) for the multi-role vessel's LARS together with its organic RHB commenced upon completion of the first vessel in 2015. The organic RHB was an 8.5m long rescue craft that could be recovered on either the port or starboard stern ramp of the multi-role vessel. Multiple modes of launching and recovery were tested, such as using the hydraulic rollers, the winch rope, and manual-override launch and recovery modes. These tests were repeated for both stern ramps under harbour conditions. The scope of HAT also encompassed installation checks, pressure tests of hydraulic systems, verification of safety interlocks and system functions. As part of the system engineering and integration (SE&I) performance verification process, HAT were conducted for every hull within the class of multi-role vessel.

4.3 SEA ACCEPTANCE TESTS

The true test of the twin stern ramp LARS design necessitates demonstration of its ability to perform under mission representative conditions. Following the system-level performance and safety checks during HAT, Sea Acceptance Tests (SAT) were conducted to put the LARS system design through its paces in the South China Sea. SAT protocol focussed on measuring the effectiveness of RHB launch and recovery operations. Qualitative comments from the RHB Coxswains were recorded on the approach and ease of alignment with the mothership, especially at higher sea-states. The test scope launch and recovery modes at HAT were repeated at sea with the multi-role vessel moving along a straight course in the open seas. Secondary test data collected during SAT comprised ambient noise level measurement to ensure that machinery noise did not adversely affect inter-crew communications during the launch and recovery operations. As part of the SE&I performance verification process, every ship within the class of multi-role vessel underwent SAT.

Figure 8. The multi-role vessel's Acceptance Test at Sea

5. DISCUSSION

5.1 MODEL TANK TESTS

The model tank tests successfully established the indicative operational envelope for RHB launch and recovery operations on the stern ramp, as well as load and acceleration inputs for further optimisation of stern ramp design. The influence of the tested wave peak period on RHB launch and recovery operations were not significant. Notable observations were the slightly easier RHB recovery operations at mothership speed of higher speed test runs. This was attributed to better RHB handling characteristics at higher speed. At higher sea-states, RHB launch and recovery operations were challenging, with the test team cancelling approaches with possible danger. Pitching of the mothership in head seas at higher sea-states proved to be challenging for safe RHB recovery. On a few occasions at higher sea-states, ramp emergence was observed to be almost as high as the RHB bow. This increased the risk of impact between the RHB and ramp.

Figure 9: Turbulent stern waves observed at speeds

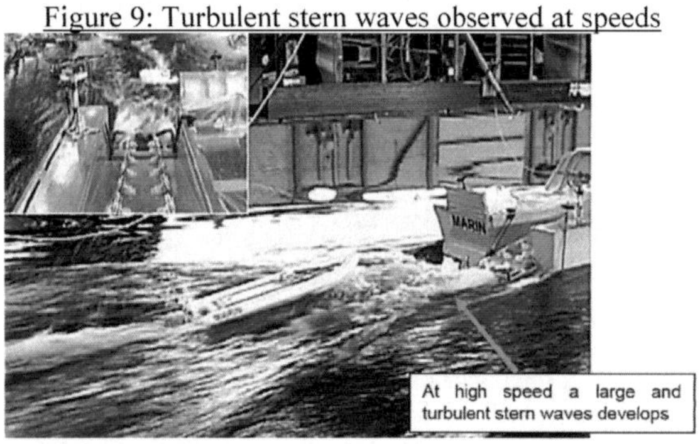

Consistent with other RHB launch and recovery tests, stern and bow quartering conditions (45- and 135 deg headings) were the least favorable headings. In higher sea-state conditions, the success of the safe stern boat operations was contingent on the helmsman's experience.

The open ramp door damped transom stern waves flow below certain mothership speeds, which aided in RHB recovery. However, in higher sea-state conditions, the vertical motions of the ramp generated by the mothership's wake made it challenging for RHB operations aft of the mothership. Lastly, test data indicated that hydraulic forces for the stern ramp door were relatively independent of the mothership speed and wave conditions. The completion of model tank tests concluded the first phase of LARS design qualification process. The team concurred with Schmittner and Carette [8] that model tests validated the selected twin stern ramp design's functionality and limits early in the design phase. This helped to avoid costly modifications to the as-built design.

5.2 TWIN STERN RAMP LARS PROTOTYPING AND INTERGRATION

The twin stern ramp LARS prototype enabled the team to further enhance its operability and safety prior to its construction on the first-of-class multi-role vessel. The following sections discuss the verification and validation (V&V) findings:

a. Hydraulic Roller Configuration. The multi-role vessel LARS hydraulic roller pairs were successfully configured to conform to hull forms of the expected range of embarked crafts, thus validating the versatility of the multi-role vessel to interoperate with various fast-response craft to support various mission profiles in a short turnaround time.

b. RHB Hull Material. Through the recovery of the various types of small craft, it was noted that craft with smoother hull materials required more time to gain sufficient traction with the hydraulic rollers which were made of marine-grade rubber. No slipping was observed during the subsequent launching of the small craft with hulls with higher coefficients of friction.

c. Human Factors. Noise and vibration levels within the craft were within acceptable ranges. This was important to ensure operator safety during RHB launch and recovery operations. The operators acknowledged the importance of the *prototyping* and *HAT* phases to refine procedures prior to the SAT and subsequent system operationalisation.

5.3 HAT RESULTS

a. Normal Launch and Recovery. Normal modes of operation include powered wheeled launching, powered wheeled recovery, and winch-assisted recovery for both the port and starboard stern ramps. For these modes of operations, launching the RHB from the multi-role vessel took less than 30 seconds. RHB recovery required less than 60 seconds – well within the designed range. This validated the multi-role vessel's ability to rapidly launch and recover RHBs, providing the multi-role vessel with a rapid-response asset for its mission spectrum. No slipping or retardation of the RHB on either stern ramp was observed during all launch and recovery tests.

b. Secondary Launch and Recovery Modes. Secondary modes of operation tested include (1) launching and recovering with hydraulic power unit crossovers from the port to starboard stern ramp and vice versa; (2) launching and recovering with one hydraulic line disabled; (3) launching and recovering with hand pumps to actuate the hydraulic rollers, and (4) emergency launching using the RHB's own weight and free-wheeling rollers to slide down the stern ramps. Secondary launch and recovery modes required longer durations to complete than the normal modes, with the exception of the emergency launch mode which took less than 15 seconds. No slipping or retardation of the RHB on the stern ramps was observed during all the launches and recoveries of the RHB.

c. RHB Alignment before Recovery. Under calm harbour conditions, the RHB Coxswain found it easy to align the RHB to the centre of the stern ramp for recovery onto the multi-role vessel. Line-up lights were installed following the prototype phase to improve guidance under low-visibility conditions.

5.4 SAT RESULTS

a. Normal Launch and Recovery. The modes tested in open sea were the same as that tested in harbour - powered wheeled launching, powered wheeled recovery, and winch-assisted recovery for both the port and starboard stern ramps. The SAT results were very similar to the HAT results, thus proving that the HAT was effective in detecting system issues upstream and ensuring that the LARS was ready for the SAT. This was especially important for sea trials where time and resources were a premium. No slipping or retardation of the RHB on the stern ramps was observed during all launch and recovery tests, despite the roll and pitch of the multi-role vessel.

b. Secondary Launch Modes. The emergency launch was tested in open sea for both the port and starboard stern ramps, which yielded similar results to the HAT. No slipping or retardation of the RHB on the stern ramps was observed during launch and recovery of the RHB, despite the roll and pitch of the mothership.

c. RHB Alignment before Recovery. The alignment of the RHB to the centre of the multi-role vessel stern ramp in open sea required a considerable amount of time due to the resultant wake of the lowered stern door, and in part due to operator familiarity with the RHB. Multiple line-up attempts were needed for each successful recovery. By analysing video footage and interviewing skilled coxswains, the team iterated with improved designs for the stern ramp fenders and guide poles to test at subsequent sea trials. Experimentation converged on a final design that resulted in shorter alignment and recovery durations. The final fender designs successfully channelled the RHB to the centre of the stern ramp even for less ideal approach angles, hence increasing the RHB recovery envelope.

d. Human Factors. Operators on board the RHB found that forces and acceleration during launch and recovery were well within safe limits for all operating modes, despite the roll and pitch of the mothership. For operators on the mothership quarterdeck, voice commands to and from the RHB personnel could be effectively heard over the ambient noise. During the winch-assisted recovery, operators experienced little difficulty retrieving the floating winch line and securing it to the RHB despite the wake caused by the multi-role vessel. Operating the LARS for both RHB and mothership crews proved to be much easier, faster, and safer as compared to davit crane operations, thus validating the effectiveness of the twin stern ramp LARS for multi-role vessels, where fast turnaround time and minimal impact to other operations were highly desired.

5.5 FUTURE DIRECTIONS

The many advantages of LARS for multi-role vessels motivate continual design improvements and incorporation of emerging technology to enhance safety and operability. The twin stern ramp LARS on board the multi-role vessel continues to serve as an effective testbed to assimilate digital technologies and new sensors for future multi-role vessels for the RSN. Prospective future directions include:

a. LARS Simulator Training. In order to reduce RHB and ship downtime for training, high fidelity models for LARS operations (encompassing both small craft and mothership motions on the sea) can be developed where crews can be trained in a virtual reality simulator. Augmented reality wearables can also be used for training coxswains for faster and safer LARS operations, with real-time information such as waypoints, steering, and water currents being piped from ship sensors to a heads-up display.

b. Operational Testing and Evaluation (OT&E). The multi-role vessel's configurable hydraulic roller pairs allow small craft of different hull forms to be launched and recovered on the stern ramps. LARS trials on future rapid-response crafts would redefine the range of interoperable mission assets. These trials have also helped to streamline procedures for deck crews, and support experimentation with new payloads. These inputs were useful in establishing the cardinal requirements and trade-offs for LARS designs on future multi-role vessels.

c. Supporting Unmanned Launch and Recovery. The proliferation of Unmanned Surface Vessels (USVs) in maritime security and search-and-rescue operations drive the impetus for future LARS to integrate sensors and communication links between the mothership and USVs for fully autonomous launch and recovery – completely eliminating the human operators from the loop. Proximity sensors at the front of the USV could augment computer vision recognition of recovery patterns and real time propulsion corrections by a shipboard stability computer. This would assist in the alignment of the USV to the mothership's stern ramp for recovery. Unmanned LARS are an attractive proposal for faster, safer operations. V&V of associated system logic remained a key area to underwrite the systems' robustness. On the motherships' end, proximity sensors could be installed to enhance the accuracy of the USV alignment algorithm and autonomous winch rope attachments could secure the rope end to the USV without human intervention. The potential manpower savings would be an attractive proposition for leanly crewed multi-role vessels.

6. CONCLUSION

The twin stern ramp LARS provides a versatile means for multi-role ships to interoperate with a wide range of embarked rapid-response vessels such as RHBs and USVs. Key benefits identified were multi-craft operability, faster turnaround time, manpower savings, and reduced safety risks as compared to traditional crane and davit launch and recovery operations. The twin stern ramp LARS design journey entailed SE&I and V&V processes such prototyping and as model tank tests, prior to acceptance tests under harbour and mission representative conditions. These processes successfully put the system through its paces for operationalization and future OT&E with other rapid-response crafts like USVs.

7. ACKNOWLEDGEMENTS

The authors acknowledge the contributions from the RSN, STEML, SAAB Kockums (then-ThyssenKrupp Marine Systems AB), and Palfinger to the successful development and integration of the twin stern ramp LARS on a multi-role vessel.

8. REFERENCES

1. Cosmo, A., "Naval Undersea Warfare Center (NSWC) – Designed Rigid Hulled Inflatable Boat (RHIB) Capture Latch Test Results" ASNE Launch and Recovery of Manned and Unmanned Vehicles from Surface Platforms Proceedings, 2008.

2. Petersen, S., "Unmanned Surface Vehicle Autonomous Recovery System Development Work at Naval Surface Warfare Center Combat Craft Division FY06-FY07" ASNE Launch and Recovery of Manned and Unmanned Vehicles from Surface Platforms Proceedings, 2008.

3. Galway, R., "Design Considerations for Autonomous Launch and Recovery of Autonomous Unmanned Surface Craft" ASNE Launch and Recovery of Manned and Unmanned Vehicles from Surface Platforms Proceedings, 2008.

4. Kilbourne, P., "All-Weather Retrieval System" ASNE Launch and Recovery of Manned and Unmanned Vehicles from Surface Platforms Proceedings, 2008.

5. Dallinga, R., E. Harmsen, "Boat Launch and Recovery from the RNLN Patrol Ship" ASNE Launch and Recovery of Manned and Unmanned Vehicles from Surface Platforms, 2008.

6. Carette, N., "Model Tests and Time Domain Simulations for Launch and Recovery Systems with Manned and Unmanned Vehicle", Launch and Recovery Symposium, 2014.

7. McTaggart, K., "Hydrodynamic Interactions during Launch and Recovery of a Small Boat from a Ship in a Seaway" Defence Research and Development Canada – Atlantic Research Centre, 2014.

8. Schmittner, C., Carette, N., "State of the Art Model Test Procedures for Launch and Recovery", Launch and Recovery, 2010.

9. AUTHORS BIOGRAPHY

Dr KOH Leong Nigel holds the current position of Head Engineering (Naval Architecture) at the Defence Science and Technology Agency (DSTA), Singapore. He is responsible for capability development in naval architecture for surface warship. His previous experience includes design and development, prototyping and model testing of the LARS for Littoral Mission Vessel. He holds a PhD in Marine Technology from University of Newcastle-upon-Tyne, UK, MSc in Maritime Studies,

Nanyang Technological University and BEng (1st Class Honours) in Small Craft Technology from University of Newcastle-upon-Tyne, UK.

David WONG Bingxiong is a Senior Engineer at the Defence Science and Technology Agency (DSTA), Singapore, where he integrates unmanned-, marine-, and mission systems on multi-role naval vessels. His experience spans naval aviation, automation design, and systems engineering & integration on the Littoral Mission Vessel. David holds a BEng (Hons) Mechanical Engineering (Design) from Nanyang Technological University, and Dip. (Merit) Biomedical Engineering from Ngee Ann Polytechnic.

TAN Yi Ming, Justin holds the current position of Senior Engineer at Defence Science and Technology Agency (DSTA), Singapore, where he integrates marine, hull and unmanned systems on board multi-role vessels. He is experienced in assessing the feasibility of launching and recovering small craft on board the Littoral Mission Vessel. Justin graduated with an MSc in Management Science & Engineering from Stanford and BSc in Mechanical Engineering from the University of Illinois, Urbana-Champaign.

CHEW Boon How holds the current position of Senior Engineer at the Defence Science and Technology Agency (DSTA), Singapore, where he integrates marine and hull systems on multi-role naval vessels. He is experienced in the design and development of LARS for the Littoral Mission Vessel and integration efforts for LMV's organic RHB. Boon How holds a BEng (Hons) Aerospace Engineering from Nanyang Technological University

Tessa GAN Su-Shan holds the current position of Director (Capability Development) at the Defence Science and Technology Agency (DSTA), Singapore. She led the team during the design and construction phase of the Littoral Mission Vessel, and was involved in the early stages of evaluation of the selection of the stern launch and recovery system as well as in defining the various tests for the validation of the concept and design. Tessa holds a MSE in Naval Architecture from University of Michigan and a BEng (Hons) in Naval Architecture from University of Newcastle Upon Tyne, UK.

THE DEVELOPMENT OF A MODULAR MULTI-ROLE VESSEL FOR NAVAL AND COASTGUARD OPERATIONS

A D Humphries, L J Davies, C P Browning and **J E Fox-Robinson,** Vectis Marine Design Limited, UK

SUMMARY

To deliver the full range of operational and support roles naval and coast guard forces typically employ many different types of specialist craft. Their specialist nature and the need for redundancy means that these craft are often under-utilised, spending much of their lives waiting for a particular requirement or incident to occur. As a result these critical and expensive assets can deteriorate through lack of use and become increasingly expensive to operate. A modular multi-role solution has been developed to address these issues. The system comprises a standard platform that can be fitted with one of a number of specialist modules. These self-contained modules can be rapidly exchanged to enable the craft to meet planned and unplanned operational and support requirements.

1. INTRODUCTION

Naval forces, coastguards and other government agencies that operate vessels are under constant pressure to reduce costs while at the same time maintain or expand their operational capabilities to meet new and emerging requirements and threats. The same applies to non-governmental operators of vessels to perform a range of commercial operations.

The capital, operating and maintenance costs of vessels of all sizes means that efforts to reduce costs inevitably focus on vessel numbers and many operators, government and commercial, are having to deliver their current and future capability with smaller fleets than previously.

Multi-role vessels are one solution to this challenge but inevitably a more flexible multi-role vessel will be less capable in any particular role than a specialist vessel dedicated to that role. Modularity can help overcome this limitation, enabling a vessel to perform a wide range of specialist duties with the same capability as a dedicated vessel.

This paper describes the development of a modular multi-role solution whereby a standard platform can be fitted with one of a wide range of modules to enable it to perform a specialist role.

2. OPERATIONAL ROLES

To deliver the full range of front line and support roles naval forces, coastguards and other operators typically employ many different types of vessel to deliver all of their required roles. These roles include;

- Military Patrol - front line patrol for security, deterrence and enforcement duties
- Non-military Patrol - second line patrol for illegal immigration and smuggling prevention, fishery protection, etc.
- Police patrol, surveillance and command
- Harbour patrol/port security
- CBRN monitoring and rescue
- Passenger transfer
- Cargo transfer
- Dive support
- Hydrographic survey
- Training
- Range safety
- Target towing/deployment
- Search and rescue
- Fire-fighting
- Ambulance
- Pollution control
- General workboat

3. SPECIALIST VESSELS

It can be seen that there are a significant number of front line and support roles that each need to be provided for. Some of these roles, such as routine patrol, training, hydrographic survey and general workboat duties can generally be planned in advance, while others require continuous short notice availability. Some of the operations that can be planned in advance may be only an intermittent requirement, such as hydrographic survey, target towing/deployment, range safety, navigation and seamanship training, dive training, and mooring maintenance.

Capabilities that require continuous short notice availability include all those required for emergency response to a marine accident, oil spill, air crash, civil emergency, natural disaster or terrorist incident. Vessels are then required for search and rescue, fire-fighting, ambulance and pollution control operations. But by their very nature such emergencies occur only intermittently.

Specialist vessels, whether for planned routine operations or for emergency response can spend a large proportion of their lives either waiting for an intermittent operational requirement or an emergency incident. As a result, specialist vessels are often significantly under-utilised.

Under-utilised vessels tend to deteriorate through lack of use and operating and maintenance personnel become less familiar with them unless regular operation and effective planned maintenance is in place. Furthermore, each type of specialist vessel has its own operating characteristics and maintenance and spares requirements so the logistic and training requirements of a fleet of dedicated specialist vessels can be significant. The specialist equipment will in some cases have a greater maintenance requirement than the vessel itself leading to reduced vessel availability even though it is otherwise fully operational.

However perhaps the greatest drawback of dedicated specialist vessels is that if the vessel suffers a breakdown the specialist capability is unavailable even if the specialist equipment itself is operational, and of course the reverse is also true. The natural solution is to hold at least two of each specialist vessel within the fleet to provide redundancy and maintain availability during planned maintenance and breakdown. The drawback of this is that the operator then has two even more under-utilised vessels to purchase and maintain each fitted with the full specialist capability.

Dedicated specialist vessels have a number of drawbacks that provide fleet operators with operational challenges and high capital and operating costs. Examples of some typical specialist vessels are illustrated below (Figures 1, 2 and 3).

Figure 1. A typical naval training vessel

Figure 2. A typical civilian ambulance boat

Figure 3. A typical commercial survey vessel

4. REQUIREMENTS FOR A MODULAR MULTI-ROLE VESSEL

The concept of the Modular Multi-Role Vessel (M2RV) was developed to overcome the drawbacks associated with a fleet of specialist vessels. The concept is based on a standard platform vessel and a standard module envelope with a set of inter-changeable modules that can be rapidly replaced to re-role the vessel when required. This concept is not new but the development of an effective solution presents numerous technical challenges that must be overcome to avoid the "jack of all trades, master of none" compromise. Also a solution that enables a vessel to be re-roled in slow time to meet planned routine operations will not provide a solution for unplanned emergency response when re-roling must be completed within the shortest possible timescale.

The requirements for the M2RV can be summarised as;

- A standard platform with common arrangement, propulsion, systems, design standards and certification
- Inter-changeable self-contained specialist modules including life-saving appliances for the specialist personnel
- Plug and play power, communication and data from the platform vessel
- Maximum speed up to 30 knots depending upon module installed
- Operating envelope to enable operation at 20 knots in Sea State 3 with any module installed.
- Up to 48 hour continuous operation up to 60 nm from a safe haven
- Ability to re-role the vessel afloat within 30 minutes
- Modules to be transportable by road and cargo aircraft
- Classification society approved platform and modules

The specialist modules include the following roles and an outline requirements specification was prepared for each of these.

- Armed patrol
- Dive Support
- Fire-fighting

- Ambulance
- Search and Rescue
- Pollution Control
- Passenger Transfer
- Hydrographic Survey
- General Workboat
- Training
- CBRN

Each specialist module being self-contained was a key requirement in order to achieve the target re-roling time but also to minimise the number of inter-connections. However, it was recognised that limitations in module size required certain services to be provided by the platform vessel. This would include electrical power and water supply, but also galley and mess facilities, sanitation facilities and overnight accommodation for the module personnel, however limited these may be.

5. CONCEPT DEVELOPMENT

5.1 GENERAL

The optimum size of the M2RV platform was identified following an iterative concept design process that considered the weight of the heaviest module and the desired footprint of the module requiring the largest internal deck space.

Inevitably on a smaller vessel the module has to be installed on top of the watertight hull envelope rather than inside it. Hence the design of the M2RV platform is driven by the weight and size of the heaviest and largest modules respectively. The Dive Support Module was identified as the heaviest due to the weight of the diving air compressor and air cylinders required. For all modules providing the largest possible internal volume and in particular deck space is desirable but the dive support, training and passenger transfer were collectively identified as requiring the most internal deck space.

5.2 MODULE DEVELOPMENT

Standard shipping containers provide an obvious option as the basis for module development. Standard shipping containers provide a number of benefits including low cost, high payload, high strength and robustness, and well established transportation, handing and securing solutions. Their drawbacks for use as the modules for smaller vessels include their high weight and limited internal volume, which is a factor when it is feasible to install only a single module.

Shipping containers are available constructed in both steel and aluminium alloy with the latter providing a worthwhile weight saving. Of course it is feasible to design and construct a module from composite materials to the same external geometry as a standard shipping container in order to save weight but this does not overcome the internal volume constraint.

To assess the feasibility of using a standard 20' ISO container a series of concept arrangements were developed. These revealed that several specialist roles could be accommodated within the envelope of a standard 20' ISO container, including the Hydrographic Survey and other roles with lower space requirements (Figure 4).

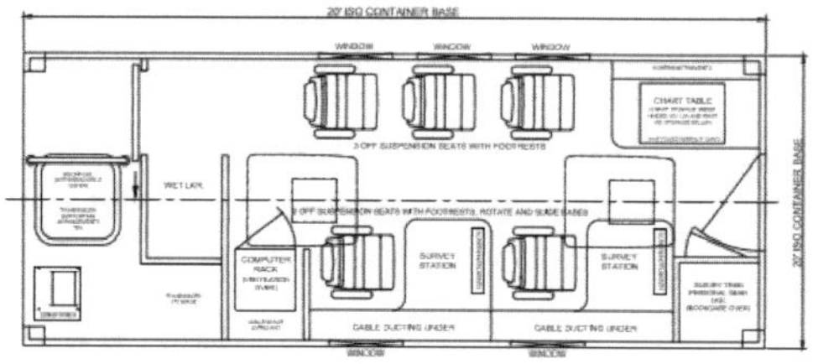

Figure 4. Hydrographic survey module within a 20' ISO container footprint

However, several other roles, such as the Dive Support, Training and Passenger Transfer could not feasibly be accommodated with the envelope of a standard 20' ISO container (Figure 5).

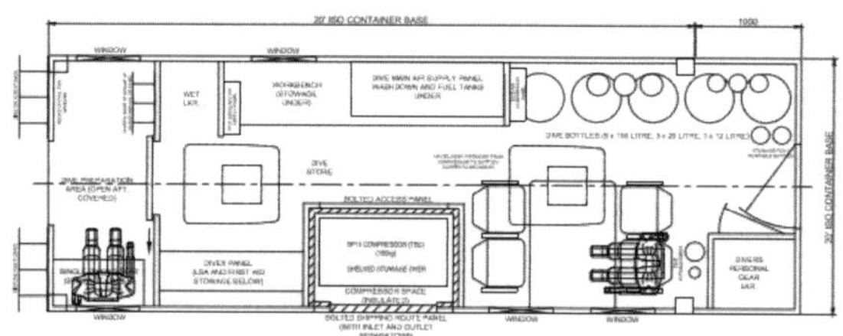

Figure 5. Dive Support module within a 20' ISO container lengthened by 1m

Clearly lengthening an ISO container by any amount immediately loses all of the benefits of the standard container, as does the mounting of any items outside the container envelope, such as liferafts, antennae and other equipment unless they can be quickly de-mounted. The same applies to the incorporation of a stowage "sump" under the module which was developed to increase the available storage volume.

In light of the space limitations of standard shipping containers the decision was taken to abandon this option and design a module to provide greater internal volume as well as reduced weight.

Various module footprints were assessed before the optimum length and width was identified. The module height is driven by the requirement to provide comfortable headroom plus the base and deck structure depth, including head linings and lights. The selected module geometry is shown in Figure 6.

Figure 6. Adopted module geometry

5.3 HULL TYPE SELECTION

The two hull types were considered for the M2RV were a semi-displacement mono hull and a conventional catamaran. The pros and cons of each were assessed in relation to the platform requirements including;

- Load carrying ability
- Transverse stability
- Deck area
- Powering requirements
- Sea keeping ability

It will be recognised that several of these factors are inter-linked, for example, a wider vessel such as a catamaran will have greater deck area and greater transverse stability. A high degree of transverse stability is desirable as the module inevitable raises the overall VCG with the associated likelihood of higher roll amplitudes when the vessel is stationary or operating at slow speeds. Similarly, a hullform with lower power requirements due to its flatter hull bottom will inevitably have reduced seakeeping ability.

The monohull scores highly against most of the requirements. The obvious drawbacks for a modular vessel are their limited beam and hence limited deck area and reduced transverse stability. Monohulls also have higher power requirements, especially when heavily loaded, but this is a trade-off with their seakeeping ability.

A conventional catamaran also scores highly against many of the criteria in particular deck area and transverse stability. One drawback is that the module and wheelhouse have to be mounted relatively high due to the limited depth of the cross-deck structure. Catamarans have good seakeeping ability up until the point when slamming of the cross-deck occurs when speed reduction is then necessary.

In summary, a monohull was considered to provide the optimum solution for those roles requiring operation at higher speed in higher sea states. For those roles requiring prolonged periods either stationary or at slow speeds in lower sea states a catamaran provides the optimum solution.

6. DESIGN DEVELOPMENT

6.1 GENERAL ARRANGEMENT

Many of the specialist roles require some type of deployment which is inevitably best carried out over the stern of the vessel. This includes divers, daughter craft, towed hydrographic survey arrays, pollution control booms, etc. and these requirements therefore support an aft module location. An aft location is also preferred for most specialist operations as the specialist personnel and equipment is located in the area of reduced motions when at sea. However an aft module location then dictates that the platform wheelhouse is located forward as the size of craft limits the scope for a raised wheelhouse that looks over the module. As a result the platform crew who will spend far greater periods of time on board are located forward in the region of increased motions and vertical accelerations leading to increased rates of fatigue. This drawback was considered unavoidable but would be mitigated by locating the wheelhouse as far aft as possible and optimising the hullform to minimise high vertical accelerations when underway at speed at sea. Figures 7 and 8 show the monohull and catamaran M2RV platform designs.

Figure 7. The monohull platform arrangement

Figure 8. The catamaran platform arrangement

6.2 HULLFORM

The hull form adopted for the M2RV monohull platform was the Vectis Advanced Planing Monohull (APM) design. The Vectis APM was developed to provide a stable and easily driven hull that combines low incidence and magnitude of slamming with low deck wetness and spray generation. It was recognised that some current wave-piercing hull designs suffer from significant spray generation when operating at speed even in calm and smooth seas. In some cases this results in frequent clouds of spray engulfing the craft and while this may have only a marginal increase in resistance it requires continuous operation of windscreen wipers to maintain visibility and means that side windows and overhead hatches which may otherwise be opened for ventilation in fine weather have to remain closed. One of the objectives of the APM was to avoid this undesirable spray formation while delivering reduced motions, slamming and accelerations with the resultant benefits for crew fatigue and safety. This has been achieved by careful bow design and combining fine forward sections with wide flatter sections aft that provide high stability and propulsion efficiency.

The APM hull form was developed and tested by scale model testing in the Southampton Solent University towing tank in calm water and in waves using a 1/18th scale model at model speeds at full-scale speeds of up to 37 knots (Figure 9).

Figure 9. Tank testing of 1/18th scale APM model

The tank testing was then simulated using Orca 3D Computational Fluid Dynamics (CFD) software prior to detailed optimisation of the hull form (Figure 10). This optimisation included refinement of the bow lines and chines to minimise and catch spray generation.

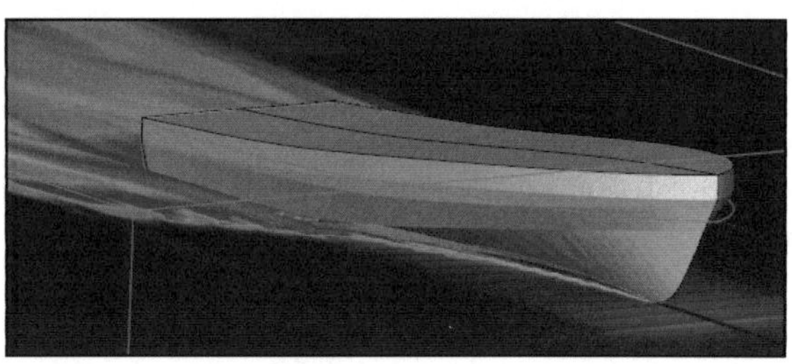

Figure 10. CFD optimisation of the APM hull form

Figure 11 shows the final APM hullform.

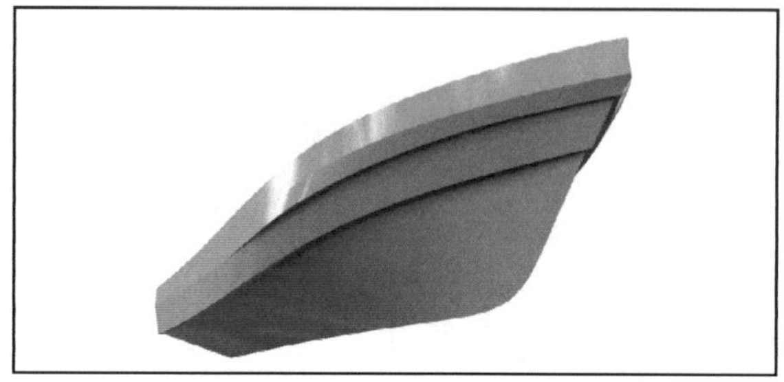

Figure 11. The Vectis Advanced Planing Monohull

The M2RV catamaran hullform adopted is based on a design proven in service on various sizes of wind farm support vessels. The symmetric catamaran design has full bow lines to minimise cross deck slamming in higher sea states.

6.3 CONSTRUCTION

To minimise propulsion power requirements the platform and module weights need to be kept to the minimum consistent with simple robust and damage resistant construction that meets Classification Society standards. Aluminium alloy and glass-reinforced plastic (GRP) construction are both obvious solutions to meet these requirements.

The M2RV platform hull, deck and wheelhouse structures were designed to Lloyds Register Special Service Craft rules. The module structures are also designed to LR SSC rules with the additional requirements for securing point structure and lifting structures. The modules were not required to be lifted with personnel on board and the design load factors were specified accordingly.

6.4 PROPULSION

Three propulsion options are applicable to the M2RV; conventional shafts with fixed pitch propellers and rudders, waterjets, and podded drives such as the Volvo Penta Integrated Propulsion System (IPS). A twin installation of either system is necessary to provide redundancy, safety and manoeuvrability requirements and the monohull and catamaran hullforms can accommodate any of the three options (Figure 12). Other propulsion packages such as petrol or diesel outboard engines and surface piercing propellers are not applicable to this type of craft.

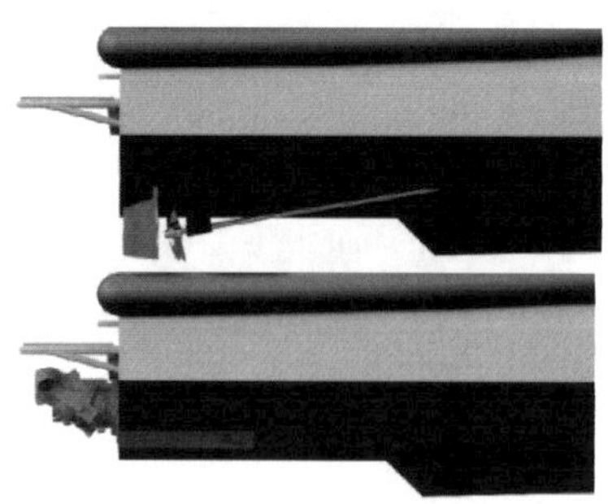

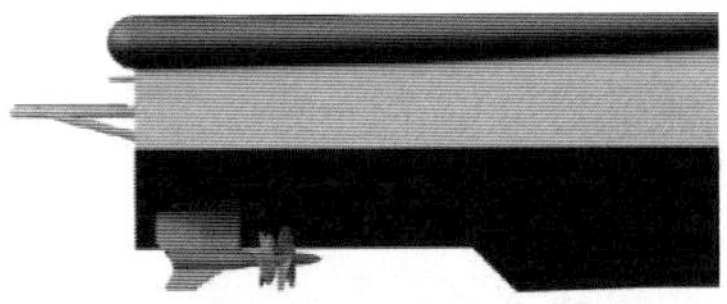

Figure 12. M2RV propulsion options; twin conventional shafts and propellers, waterjets, and Volvo Penta IPS

Each propulsion option offers advantages and drawbacks and it is largely a matter of operator preference that drives the selection. Waterjets provide excellent manoeuvrability combined with shallow draft and reduced vulnerability to damage, as well as greater safety to divers and personnel in the water, albeit at higher capital and maintenance cost and lower efficiency. Conventional propellers and shafts provide an efficient, cost effective and low maintenance solution but with a deeper draft. Podded drives offer higher efficiency and high manoeuvrability but also with greater draft and vulnerability to damage, and higher initial cost.

6.5 STABILITY

The intact and damaged stability were analysed with the heaviest module installed to the UK Maritime and Coastguard Agency Workboat Code and meets all of the criteria with suitable margins. In addition, the vessel had to meet the IMO High Speed Craft code due to the requirement to carry more than 12 passengers at more than 20 knots. The HSC code is an onerous standard for any small vessel as damage scenarios include bow, stern and side damage and hull bottom damage due to raking. To meet this requirement the hull is divided into five watertight compartments and the permeability of some compartments reduced by additional buoyancy material. Figure 13 shows the floatation with the three forward compartments flooded due to raking damage.

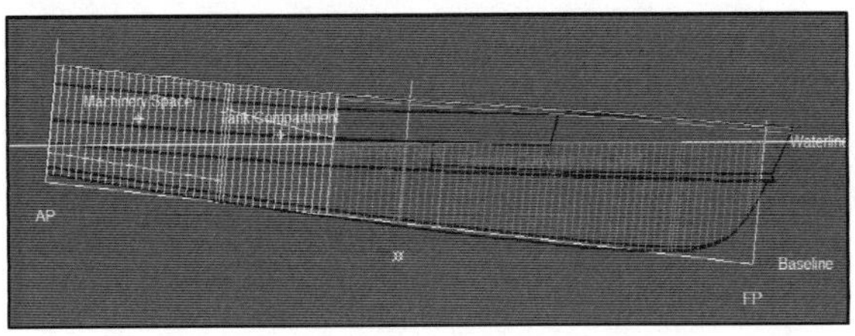

Figure 13. The vessel with the forward three compartments flooded in accordance with the IMO HSC code raking damage criteria

The damaged stability requirement illustrates how a modular multi-role vessel has to meet a number of different design standards and certification requirements applicable to the range of specialist modules and the number of operators and passengers associated with those modules. This also applies to numerous other aspects in addition to the stability requirements.

6.6 MODULE DESIGN

A common structural design was developed for all modules based on the dive support module as the heaviest module. The structural scantlings could have been reduced for lighter modules but this would add design and manufacturing complexity which the aim was to avoid.

The securing method was developed based on the securing method for standard ISO containers to provide a rapid securing method that is well-proven and Classification Society approved. Alternative securing solutions such as bolting were rejected as being unlikely to meet the rapid re-roling requirement.

As previously mentioned, one of the challenges of modular small vessels is the raised overall VCG due to the module being installed on top of the hull envelope. This is particularly an issue with the heaviest modules and a "sump" was incorporated into the module design to address this. This sump is an integral part of the module structure and provides additional stowage space but its primary purpose is to house the heaviest specialist equipment at a lower VCG. For example, the sump of the Dive Support module houses the diving air bottles which collectively weigh several hundred kilograms. The sump inserts into a recess in the deck of the M2RV platform below the module deck (Figure 14 and 15).

Figure 14. The Dive Support module showing the diving air bottle stowage that is integral with the module

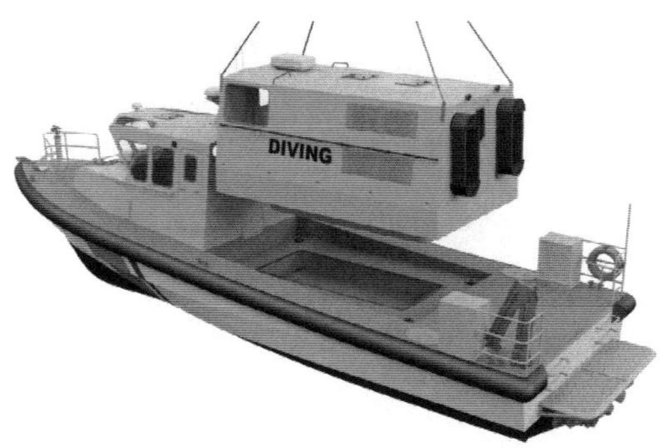

Figure 15. The M2RV and Dive Support Module

The Dive Support module also houses a diesel engine driven diving air compressor and its systems including fuel tank, air vents and fire suppression system, and divex panel (Figure 16).

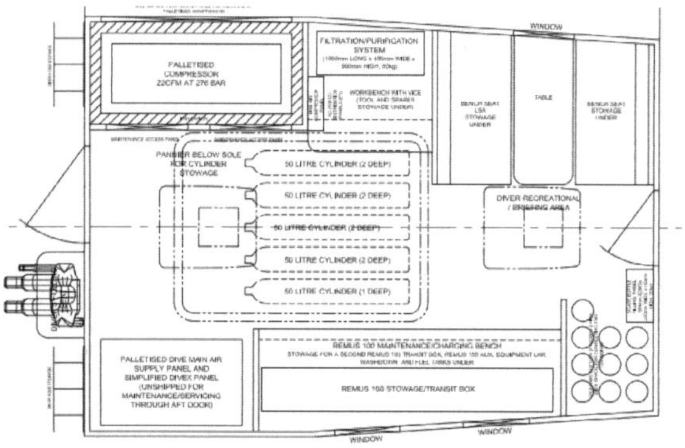

Figure 16. Internal arrangement of the Dive Support module

The requirement for each module to be fully self-contained required the development of several innovative solutions. The most significant of these was for the Hydrographic Survey module which necessarily requires connection to survey transducers mounted underwater. On monohull survey vessels these transducers tend to be permanently mounted in a fairing fitted to the hull, while on catamaran survey vessels the transducers are typically mounted on a retractable structure that deploys them through a moonpool opening through the cross deck structure.

Mounting the transducers in the hull would mean that they would be permanently installed regardless of the vessel's role, but more significantly it would require their data connection to the Hydrographic Survey module whenever it is installed, and the opposite when removed. Not only would this increase the re-roling time but it would also separate key elements of the specialist equipment with the associated maintenance and logistic challenges.

To avoid this, the transducers are mounted on a fairing deployed by a telescopic structure from within the module through a moonpool in the hull of the monohull or the cross-deck of the catamaran. This is shown in Figures 17, 18 and 19. This solution requires a plug to be installed to close the moonpool when other modules are installed.

Figure 17. The Hydrographic Survey module showing the transducer fairing deployed

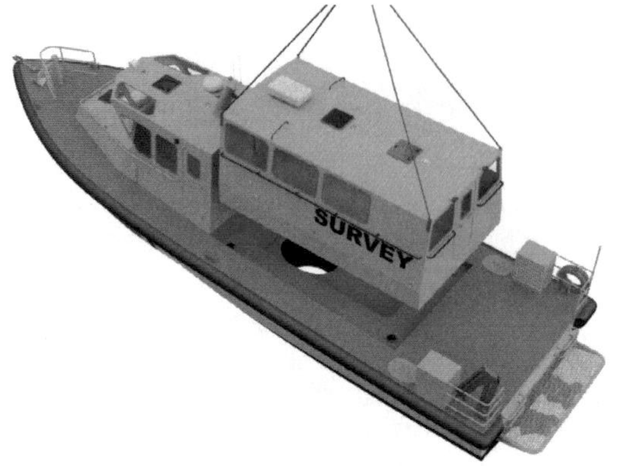

Figure 18. Installing the Hydrographic Survey module on the M2RV monohull platform

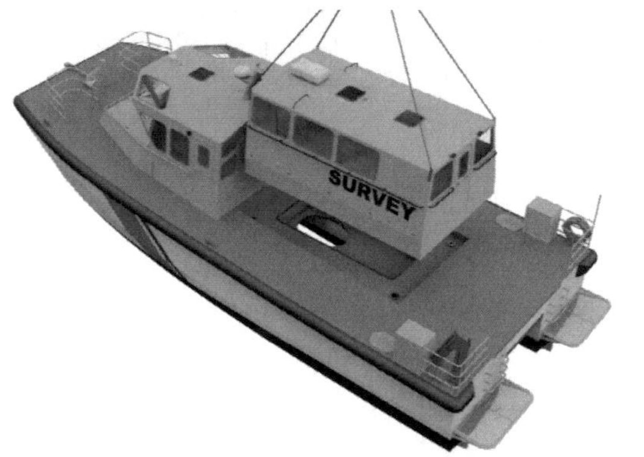

Figure 19. Installing the Hydrographic Survey module on the M2RV catamaran platform

All of the modules rely on the platform vessel for electrical power, standard intercom and data links, fresh water supply and chilled water for air-conditioning. In addition, the fire-fighting module relies on a sea water supply to the fire monitor from the platform vessel. All of these are made by quick-release connections. Each module is also fitted with all of the life-saving appliances for the operators and passengers associated with that module. Some of the specialist modules are illustrated below. Figure 20 shows the Armed Patrol module which is fitted with remote weapon station and surveillance radar. It can be seen that one of the compromises of the aft module location is the restricted forward arc of fire.

Figure 20. Armed Patrol module

The fire-fighting module is shown in Figure 21. The fire monitors are mounted aft and have limited forward operating ability.

Figure 21. Fire-fighting module

Figure 22 shows the Pollution Control module which is fitted for oil spill first response with an oil containment boom and detergent spray booms. The detergent is stored in a tank within the module "sump". The ability of the vessel to store recovered oil is limited and it is envisaged that this role will be provided by a larger vessel.

Figure 22. Pollution Control module

Figure 23 shows the Police Surveillance/Forward Command Centre module for civilian police operations.

Figure 23. Police Surveillance/Forward Command Centre module

Figure 24 shows the General Workboat module equipped for mooring and navaid maintenance, salvage, towing and other general workboat duties. The A-frame and the optional small RIB or ROV are examples of items that could not be readily incorporated within the adopted module envelope. Extending the module base to the transom was considered but it would have led to a much larger module that was not required by the majority of the specialist roles.

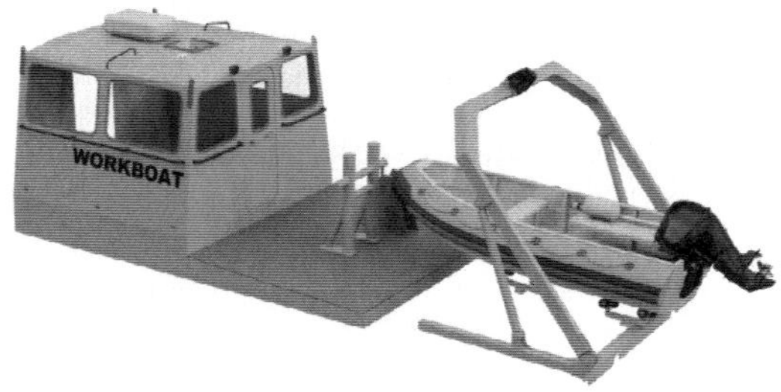

Figure 24. General Workboat module

Figure 25 shows the monohull M2RV with the Armed Patrol module installed.

Figure 25. The monohull M2RV with Armed Patrol module installed

7. ADVANTAGES AND DRAWBACKS

7.1 SPECIALIST VESSELS

7.1.1 Advantages

The most obvious advantage of dedicated specialist vessels is that their design can be fully optimised for their single role. In theory this can be without compromise, but in reality this is unlikely to be the case as specialist vessels are only rarely designed from scratch and more often the design is based on a parent vessel design.

7.1.2 Drawbacks

The drawbacks of specialist vessels can be summarised as follows.

- A large number of different types of specialist vessel are required to meet the full range of operational requirements
- At least two of each type of specialist vessel is required to provide redundancy
- The operator training, maintenance and logistic support requirements of a large mixed fleet is high
- Specialist vessels are often significantly under-utilised

- Under-utilisation often reduces reliability, increases maintenance requirements and reduced operator familiarity with the vessel

All of these drawbacks can lead to reduced operational capability and increased costs.

7.2 MODULAR MULTI-ROLE VESSEL

7.2.1 Advantages

The benefits of the modular multi-role vessel can be summarised as follows.

- A smaller number of platform vessels are required that are each more fully utilised
- The smaller fleet of standard platform vessels will reduced training, maintenance and logistic support requirements
- The inventory of specialist modules can be specified to meet operational requirements
- Platform vessels and modules may be procured together and/or separately to meet operational requirements and budgetary constraints.
- Modules can be purchased at a later date if a new capability is identified without having to purchase a new vessel.
- Modules can be replaced or upgraded to future-proof the specialist capability independently from the platform vessel.
- Platform vessels can be replaced or upgraded without replacing modules.
- Platform and module availability are de-coupled and redundancy is provided by two or more specialist modules rather than two or more specialist vessels.
- Platform and module maintenance are de-coupled.
- Modules can be maintained ashore in workshop conditions or returned to the specialist equipment supplier for maintenance and upgrade.

Overall the benefits to fleet operators are considered to be increased operational availability, reduced capital expenditure, greater utilization of assets, reduced operator and maintainer training, and reduced logistic support requirements. The magnitude of the benefits will depend upon the operator, the size of their current fleet and the range of duties they are required to deliver. An operational analysis study will identify the number of platform vessels and the numbers and types of specialist modules required.

7.2.2 Drawbacks

The drawbacks of the concept can be summarised as follow.

- Inevitably on a smaller vessel the module has to be located aft of the wheelhouse. Not only does this force the wheelhouse forward into the region of the highest motions, but also limits the forward arcs of any armament and equipment mounted on the module.
- Maximising the self-sufficiency of each module which is necessary for rapid re-roling does introduce space constraints and VCG challenges that can be avoided with dedicated specialist vessels where some of the specialist equipment can be mounted within the hull.
- The standard platform has to meet the design standards and certification requirements of the most onerous specialist module and the operator/passenger numbers associated with that module.
- The platform vessel has to incorporate the connections for all available modules.
- The requirement for rapid re-roling necessitates the use of quick connection couplings for power supplies and other services from the platforrn vessel. Such connections require careful design, protection and maintenance to maintain their reliability.
- As is the case with dedicated specialist vessels, some roles favour a monohull and others a catamaran platform, and the same applies to the modular multi-role vessel. The ideal fleet solution would be made up of a mix of monohull and catamaran platforms which while being fully inter-changeable would be selected whenever possible according to the role. This would reduce the degree of commonality across the fleet but this may be seen as a worthwhile compromise by some operators.

7. CONCLUSIONS

This paper has identified the advantages and drawbacks of dedicated specialist vessels that are typically operated by naval and coast guard forces, by other government agencies and commercial operators.

An alternative solution of a standard platform capable of deploying a wide range of specialist modules has been developed which offers benefits to fleet operators. The development of the solution presents a number of technical challenges to address the impact of a module on the platform design. Some of these challenges are related to the requirement that each specialist module should be self-contained requiring only the provision of electrical and other services from the platform. This objective is necessary to meet the requirement for emergency response where the ability to rapidly re-role the vessel is essential. This objective has been achieved to a large degree with only the transom mounted items of some roles being separate from the module itself.

The Modular Multi-Role Vessel (M2RV) and a range of inter-changeable specialist modules to meet the front-line and secondary roles of naval and coast guard forces has been developed. The M2RV overcomes the fundamental drawbacks of a fleet of dedicated specialist craft with minimal impact on capability. The key benefits are a reduction in the numbers of vessels required coupled with the enhanced availability and reduced costs bought about by de-coupling the specialist capability from the platform vessel. Capital and operating cost savings are predicted.

The M2RV concept offers the opportunity for a number of government agencies to co-operate to reduce costs by sharing a fleet of M2RV platforms with each agency having its own inventory of the specialist modules it requires. Such an inter-agency co-operation will deliver greater operational and cost benefits.

8. ACKNOWLEDGEMENTS

The authors are grateful to the Directors and Managers of Berthon Boat Company, Lymington, UK for their support and input.

9. AUTHORS' BIOGRAPHIES

Andrew Humphries is Founder and Managing Director of Vectis Marine Design. He is responsible for overall management and strategy and provides technical direction and input to projects. He was previously Design Manager at VT Halmatic and held various technical and management roles with Vosper Thornycroft Shipbuilding.

Les Davies is a Senior Naval Architect with Vectis Marine Design. He is responsible for all aspects of the naval architecture and design on a wide range of projects. His previous experience includes naval architecture and design management with several yacht and boat building companies including VT Halmatic.

Chris Browning is a Naval Architect with Vectis Marine Design responsible for small craft naval architecture and design with particular focus on the design of high speed craft. He joined Vectis in 2011 directly from Southampton Solent University.

John Fox-Robinson is Technical Director at Vectis Marine Design responsible for overall technical management and governance as well as providing technical direction and input to projects. His previous experience was with BAE Systems Small Craft division, VT Halmatic and WS Atkins.

STANDARDISATION OF NAVAL CONTAINERISED MODULES - ENABLING INTEROPERABILITY ACROSS NATO

H K Cole and **J Alvarez-Mari,** MOD Naval Design Partnering, UK

SUMMARY

This paper explores the work carried out by the Naval Design Partnering (NDP) team to propose a new North Atlantic Treaty Organisation (NATO) standard covering naval containerised modules outfitted with mission systems. The purpose of the Standard is to facilitate interoperability of naval modules across NATO navies, and to future proof the design process both for supply-chain and vessel designers. Drawing upon the experience of the offshore world in using containerised modules, as embodied in the current certification standards, the Standard provides a uniform set of safety-related construction requirements, standardised interfaces, and guidance for platforms to enable the deployment of naval modules. The paper covers the overall philosophy behind the Standard, the background research conducted, and the rationale behind the requirements chosen. Finally, it explores the impact of the Standard on platform design, the key enablers for the deployment of naval modules, and the future work to develop international inter-operability.

NOMENCLATURE

AECTP	Allied Environmental Conditions and Test Publications
ANEP	Allied Naval Engineering Publication
ATP	Allied Tactical Publication
AWACS	Airborne Warning and Control System
BSP	British Standard Pipe
DNV	Det Norske Veritas
EMC	Electromagnetic Compatibility
FSS	Fire Safety Systems
GbE	Gigabit Ethernet
HVAC	Heating, Ventilation and Air Conditioning
IEC	International Electrotechnical Commission
IMO	International Maritime Organisation
ISO	International Standards Organisation
MOD	Ministry of Defence (UK)
MSC	Maritime Safety Committee
NAG	Naval Authority Group (UK MOD)
NATO	North Atlantic Treaty Organisation
NDP	Naval Design Partnering (UK MOD)
NIAG	NATO Industry Advisory Groups
QPS	Quality of Power Supply
SOLAS	Safety of Life at Sea
STANAG	STANdardisation AGreement
WLL	Working Load Limit

1. INTRODUCTION

A number of work-streams across the North Atlantic Treaty Organisation (NATO) are looking at the question of mission modularity, with particular emphasis on the sharing of assets across NATO [1]. Various NATO Specialist Teams, and NATO Industry Advisory Groups (NIAG) have analysed the problem from a number of angles, including the operational analysis perspective, whole life costing, and survivability.

An analysis of the mission modularity problem would not be complete without considering the question of the requirements for module design. The Naval Design Partnering (NDP) team was tasked with addressing this aspect of the mission modularity question. This work has resulted in the production of a draft Allied Naval Engineering Publication (ANEP) on naval containerised modules (referred to throughout as "the Standard"), with the ultimate aim of it becoming endorsed by a NATO Standardisation Agreement (STANAG). The underlying philosophy for the Standard, and proposed requirements, are explored in the present paper.

2. OVERALL AIMS AND PHILOSOPHY

2.1 GENERAL

The Standard aims to provide a minimum set of requirements to facilitate interoperability across NATO navies, by providing a uniform set of safety related construction requirements, a set of standardised interfaces, and guidance applicable to the construction of new platforms to enable the deployment of modules.

2.2 THE DEFINITION OF A NAVAL CONTAINERISED MODULE

The Standard defines a naval containerised module as "a unit built and equipped for a special service task, mainly for temporary installation, intended to be operated on board a ship, in support of delivering a military effect".

The Standard only covers modules that need to operate within a container, and it has been assumed that, for ease of transport, this means the envelope of an International Standards Organisation (ISO) container, within the dimensional constraints of ISO 668 "Freight Containers - Classification, dimensions and ratings". The Standard anticipates that most naval modules would be 20' long, or more rarely 40' long.

2.3 OWNERSHIP MODEL

The Standard implicitly assumes a national model of ownership for the naval modules. Even though the aim is to facilitate interoperability, the underlying assumption is that the modules will be built for the national interest in response to national operational requirements, and that they may then be deployed on NATO missions as required. The reason for this is simply that, while shared ownership is an aspect that a number of the modularity work-streams are considering, actual shared assets are in fact very few – it is understood by the authors that the main NATO assets with truly shared ownership are the Airborne Warning and Control System (AWACS) planes and the Submarine Rescue System.

The consequence of the implicit national ownership model is that the Standard cannot be overly prescriptive. As such, it focuses on safety requirements and on the production of information packs that would enable a Naval Administration to assess in a straightforward manner whether a module of another NATO navy can be safely embarked in a national platform. In a similar manner, the interfaces are defined simply by the applicable standards. In some cases, additional detail is included for practical reasons – for example, a minimum size is recommended for fluid interfaces to facilitate mechanical handling of connections and mitigate the risk of damage.

2.4 COMPLIANCE ASSESSMENT AND THE NAVAL ADMINISTRATION

The Standard places requirements on both module and host platform, for example; provision of life jackets and emergency breathing devices for extra embarked crew, and assessment of approved module layouts for mission bays to ensure safe and clash free embarkation. The responsibility of assessing compliance against these requirements is placed upon the national Naval Administration responsible for the relevant host platform. Naval Administration is defined as "The Department of Government of the State responsible for providing safety regulation for naval ships", for example the Naval Authority Group (NAG) for UK MOD.

To enable flexibility and prevent over-prescription, the Standard empowers designers to review the expected applications for a module, and select values for external loads or constraints – this is relevant for; climatic conditions, electromagnetic conditions, acoustic environments, induced loads, shock and blast, explosive environments etc. For example, where a navy has a need to embark modules on a weather-deck then the modules will be designed for sea loading, whereas if a vessel is equipped with a mission-bay then this requirement can be relaxed. In each case, the most onerous expected requirement shall be specified. The Naval Administration has a key role in assessing compliance, endorsing deviations, and ensuring that any limitations of use are recorded in the module information package, as this will enable safe interoperability.

Appropriate information must be provided to enable an initial assessment, and subsequent re-assessment when a module is to be deployed on an alternative vessel (e.g. of another nation). The module information package is defined to contain this information, which must be configured through the life of the module (see Section 3.11). The authors recognise the challenge of keeping such information configured and up-to-date, especially across a "class" of naval modules which may be distributed globally in role. However, these difficulties are similar to, and certainly no greater than, those for maintaining configuration of naval platforms, and thus the authors are confident that Naval Administrations will determine policies to achieve this.

2.5 REGULATORY BACKGROUND

2.5 (a) ANEP-77 – The Naval Ship Code

ANEP-77 [2] is the common NATO reference point for naval ship design. One of the core principles of the Standard is that a naval module, while in operation on board a ship, is not functionally different from a compartment containing the same equipment/material, and the same regulations from ANEP-77 should apply.

2.5 (b) DNV 2.7-2 – Offshore Service Modules

The offshore industry has considerable background on the use of modules on offshore platforms, and the certification rules from Det Norske Veritas (DNV) are widely applied. In this context, it was understood that DNV 2.7-2 [3], which contains the requirements for the certification of offshore service modules (which are the conceptual equivalent to naval modules in the offshore industry), would provide a good framework for the Standard.

2.5 (c) Other Standards

NATO standards were primarily selected, but where these did not exist, other relevant international standards were chosen.

3. NAVAL MODULE REQUIREMENTS

3.1 ENVIRONMENTAL REQUIREMENTS

STANAG-4370 endorses a number of Allied Environmental Conditions and Test Publications (AECTP) applicable to NATO military equipment, which are to be applied if national standards are not more onerous. These include:

- AECTP-230 for climatic conditions
- AECTP-240 for the mechanical environment
- AECTP-250 for the electrical and electromagnetic environmental conditions

The acoustic environment is covered by a separate standard. STANAG-4293 should apply if applicable national standards are not more onerous.

There is no NATO-wide guidance on shock requirements, therefore the Standard states that the naval modules are to be assessed in accordance with relevant national guidance for the intended host platform(s) to determine whether shock requirements are applicable. If this is the case, a shock mounting solution, known as the 'pumpkin mount' system has already been developed for naval modules and is intended to be used NATO-wide. An alternative arrangement could be to shock-mount the equipment internally, noting that this may reduce useable space within the module.

3.2 STRUCTURAL REQUIREMENTS

The structural requirements are principally drawn from DNV 2.7-2 Offshore Service Modules, with some specific deviations and amendments made where additional clarity for the naval use-case is required. DNV 2.7-2 requires a container to conform to the ISO standards (overall dimensions, twist-lock connectors, stacking etc.), and to be strengthened to withstand particular load cases – particularly lifting at sea. The suitability of DNV 2.7-2 as a baseline standard was principally established from its proven usage throughout the offshore industry, the clear, qualitative parallels in usage between offshore modules and naval modules, the potential requirement to enable lifting of the modules at sea (whereas an ISO container is unlikely to be suitable); and the increased strength benefits afforded over ISO containers, particularly to withstand external induced loading.

DNV 2.7-2 structural requirements enable lifting of modules while at sea. While there is no current requirement anticipated, in future naval modules may be required to be lifted while at sea; whether aboard a single vessel for logistical movements in a mission bay, interchange of capability etc., or potentially between vessels for resupply, interoperability etc. Thus, use of DNV 2.7-2 future-proofs the Standard. To afford this capability, a module should be certified either as an Offshore Container to DNV 2.7-1, or as a Portable Offshore Unit to DNV 2.7-3. It is noted that Ship Type Service Containers to DNV 2.7-2 do not meet the requirements.

To facilitate lifting, the Standard specifies forklift pockets, and pad eyes fitted with permanently attached sling sets rated to the working load limit (WLL) (see Figure 1). The latter may drive deckhead height requirements for mission bays to achieve suitable strop angles.

Figure 1 – Offshore containers fitted with ISO pockets, and permanently attached slings © Ocean Vantage Group

Shock and blast events place significant loads on structures. Shock-load testing on ISO containers demonstrated that the 'pumpkin mount' system provided shock protection to ISO containers loaded up to 10 tonnes, although the containers displayed minor damage, particularly distortion to the transverse floor beams.

Previous work by the NDP indicated that increasing module wall thickness from 1.6 mm (to meet the loading requirements of ISO 1496-1) up to 4 mm (to meet DNV 2.7-2) could potentially provide the required withstand to the anticipated blast loads. This improves confidence in the use of DNV 2.7-2 as the basis for structural requirements for the Standard, but it is recommended that further testing of an offshore container against shock/blast events is conducted for validation.

It is noted that an offshore container, with 4 mm wall thickness, has a tare (un-laden) weight approximately double that of a standard 20' container (~ 4 tonnes, compared to ~ 2 tonnes), which would result in a reduction of the loading capacity of approximately 2 tonnes. This is not anticipated to overly constrain naval applications, both in terms of the maximum gross weight and the payloads as already tested under shock.

3.3 TRANSPORTATION REQUIREMENTS

The Standard recognises that deployment of modules will likely be via land or air transport. The normal ISO container requirements provide suitable capability for land transport. The Standard notes that for air transport, the requirements of Allied Tactical Publication (ATP) 3.3.4.1 Chapter 7 Annex D should be met.

The Standard also notes the specific challenges of loading ISO containers into some smaller transport aircraft where there is minimal spatial margin (e.g. C-130 Hercules). In these cases, any protrusions from the standard ISO envelope are likely to erode this margin and may preclude air transport. Further work is required to develop the Standard in this area.

3.4 PASSIVE FIRE PROTECTION REQUIREMENTS

Following the principle that a naval module is functionally identical with a compartment while in operation, the requirements for passive fire protection that apply should be those for a compartment of the same function, as contained in ANEP-77 Chapter VI Regulation 8 Fire Protection. These requirements closely follow the structure and content of Safety of Life at Sea (SOLAS) Chapter II-2 Regulation 9 Containment of Fire, where the passive fire protection requirements depend not only on the type of compartments on each side of a boundary, but also on the type of ship. In ANEP-77, ships are characterised as Type A, B, or C depending on crew size, with a Type A platform (number of embarked persons of 240 or greater, or which will foreseeably carry greater than 36 passengers) having the most onerous requirements.

Passive fire protection is therefore to be chosen as if for a ship compartment of the same function, with the adjacent compartment classed as Open Deck, or Special Purpose Space as applicable. From the above, it follows that a module designed to be deployed in a Type A platform would also be deployable in the smaller ones. However, because of the assumed principle that the modules are built for national operational requirements, the standard has not mandated insulation for a Type A platform. Instead, the Standard mandates that the type of platform upon which the module is intended to be deployed shall be clearly stated on the information plate and in the information pack (i.e. A, B or C as per ANEP-77).

Arrangements for openings and penetrations are to comply with the requirements of ANEP-77 Chapter VI Regulation 8, and are to maintain the fire rating of the boundary in question.

Requirements for materials to limit fire growth potential, smoke generation and toxicity have been chosen to follow ANEP-77 Chapter VI Regulations 4 and 5 where applicable.

3.5 FIRE RISK CATEGORISATION

The modules will need to be subject to a fire risk assessment, to determine the requirements for fire detection and firefighting. As guidance the fire risk categories, and associated fire detection and firefighting requirements are offered in the Standard, as shown in Figure 2 (see end).

The standard places additional fire detection and firefighting requirements on modules which are to be 'routinely manned' (defined in the Standard as "Manned for more than 2 hours in every 24-hour period, or required to be manned during an emergency or action stations"). This is to reflect that while some modules may not pose a fire risk, the manning level drives additional requirements for personnel safety.

Equipment for fire detection and firefighting is to comply with the FSS Code. The number of portable fire extinguishers is to follow ANEP-77-Chapter VI, Regulation 9.11 Arrangement of Fire Extinguishers, which in turn refers to International Maritime Organisation (IMO) Maritime Safety Committee (MSC)-Circ.1275 "Unified Interpretation of SOLAS Chapter II-2 on the Number and Arrangement of Portable Fire Extinguishers on Board Ships".

3.6 FIRE DETECTION AND ALARM REQUIREMENTS

In this domain, the Standard is divided in two sections, one applicable to all fire detection and fire alarm systems, and the second applicable only to systems integrated with the overall ship-wide fire detection and fire alarm systems. The Standard defines the required actions for the fire detection and alarm systems as follows:

- The fire alarm system shall generate audible and visible signals in and around the module where persons may be present.
- Where fitted, mechanical ventilation to the module shall stop, and fire dampers shall close automatically.

The above follows DNV 2.7-2, Section 4.7.2.1 Fire Detection and Alarm, requirements for modules without integral fire systems.

From the above an important point can be inferred: the practice of having non-integrated fire systems is acceptable in an offshore environment, where the risk of explosive atmospheres is highly prevalent and where comparatively few people would be on watch routinely. While this is not standard practice on a ship, it was considered that it could be acceptable for low risk modules. The risk assessment for a particular module would, in any case, override this recommendation if necessary.

The Standard has a requirement for a minimum of two sources of power for the modules' fire alarm system, which follows ANEP-77 Chapter VI, Regulation 7.31 Sources of Power Supply. In order to simplify the installation in spaces dedicated to modules and minimise the impact on the platform, the second power supply is specified as a battery, and the duration defined by the requirements for battery backups in DNV 2.7-2 Section 4.7.2.2.

Alarm requirements are defined as follows, in in line with DNV 2.7-2 Section 4.9 and ANEP-77, Regulations 7.24 and 7.25:

- As a minimum, an audible alarm shall be fitted internal and external to the module.

- As a minimum, a visual alarm indicator shall be fitted external to the module. This indicator shall be labelled "Fire Alarm".
- The alarm units shall be silence-able (to be activated via key-switch or other similar control to prevent unauthorised use).

3.6 (a) Supplementary Requirements for Integrated Fire Alarm Systems

The Standard defines the interaction with the ship's fire alarm control system as follows:

- The module's integrated fire alarm system shall behave as an extension to the ship's fire alarm control system.
- A module's integrated fire alarm system shall interface with the ship's fire alarm control system, providing indication upon the ship's fire detection and fire alarm control panel / indicating units that at least one detector/manual call point has alarmed in the module.
- Each module shall be capable of being individually identifiable by the ship's fire detection and alarm control system.
- The ship's fire alarm control system shall give a visual and audible warning if the signal to the module integrated alarm is lost.
- If the signal to the ship's fire alarm control system is lost, the module alarm shall activate.

It should be noted that for smaller ships (length less than 50 m), ANEP-77 does not require remotely and individually identifiable fire detectors (Regulation 7.32). Any space dedicated to carry modules would need to receive special consideration in this context.

The Standard specifies that routinely manned modules shall have a minimum of one manual call point, mounted in a suitable position near the main exit (not for escape panels), and labelled "Fire Alarm".

3.7 FIRE FIGHTING REQUIREMENTS

This section provides the requirements for portable fire extinguishers and fixed firefighting systems. The Standard also includes a requirement to have drain points in the deck, to allow draining of water from firefighting or wash down post firefighting, in line with ANEP-77 Chapter VII, Regulation 9.15.2, which calls for "arrangements for the drainage of water discharged when the (firefighting) system is activated".

3.7 (a) Portable Extinguishers

MSC-Circ 1275 notwithstanding, the standard establishes a minimum of one portable extinguisher in all cases, and two extinguishers for modules which also have an emergency exit (unless this is an escape panel). Minimum capacities for fire extinguishers have been modelled after offshore requirements according to DNV 2.7-2.

3.7 (b) Fixed Extinguishing Systems

Accepted fixed fire extinguishing systems shall be as per ANEP-77 Chapter VI-Regulation 9.13 Types of Fixed Fire-Extinguishing Systems. The hazards arising from the fixed fire extinguishing system are to be included in the module information package.

For a fixed fire extinguishing system, the standard makes no recommendation for either an on-module system, or a connection to the ship's fixed fire extinguishing system. If the latter is the adopted solution, if follows that the ship's fixed fire extinguishing medium in the area where the modules are to be located will need to be compatible with the types of fire assessed as likely to occur in the module. Information on the assumed fire extinguishing medium shall feature on the module marking plate and the module information package.

The Standard defines the required actions for the fixed fire-extinguishing system (whether automatically or manually released) as follows, in accordance with the requirements of DNV 2.7-2, Section 4.9.2:

- Where fitted, mechanical ventilation to the module shall stop, and fire dampers shall close automatically.
- The system shall signal the release to the central control station.
- Where the fire-extinguishing system (or media) may cause danger or harm to personnel, the system shall generate audible and visible alarms in and around the module where persons may be present. This alarm shall be different and distinguishable from other signals.

It is expected that the default mode of operation shall be manual, with automatic operation only in high risk cases based upon the fire risk assessment, and if the fire extinguishing medium presents no danger to personnel (e.g. water drench).

3.8 ESCAPE AND EVACUATION REQUIREMENTS

It has been assumed that for a platform intended to carry modules, a number of module arrangements would have been included in the platforms' Escape and Evacuation Analysis and Demonstration. These are defined in the Standard as 'Approved Layouts'. Modules that have not been assessed as part of an Approved Layout are deemed to conform to an Approved Layout, and are effectively interchangeable, provided they introduce no additional risk, defined as:

- Fire risk assessed as lesser or equivalent to an approved module (as discussed in Section 3.5).
- Greater or equivalent fire detection and firefighting measures

- Escape and evacuation arrangements are not affected

In order to ensure that escape and evacuation arrangements are repeatable for different module configurations, the Standard specifies the following requirements:

- Normal access or exit to be on the narrow side of the naval module unless technically unfeasible.
- As per DNV 2.7-2, modules with an internal area exceeding 20 m^2, or the distance to the main exit door exceeding 5 m, shall have an additional emergency exit. This is to be on the narrow side in an easily accessible position.
- Modules with an internal area of less than 20 m^2 are to have an escape panel fitted to the primary access door or upon a narrow side of the module in an easily accessible position.
- Doors, unless specifically stated otherwise, shall open in the direction of escape.

In practice, this requirement means that a 40' module will inevitably require an emergency exit, and that a 20' module may require one, depending on the internal layout.

Requirements for separate emergency exits are aligned in ANEP-77 and DNV 2.7-2, but DNV 2.7-2 is slightly more onerous, and has been adopted for the Standard. (ANEP-77 Chapter VII Regulation 16.3 requires an additional emergency exit unless the travel distance in the compartment is less than 7 m, with 5 m for machinery spaces).

Requirements related to the securing and locking of doors, signage and clearances internal to the module for escape and evacuation purposes, including the manoeuvring of stretchers, have been chosen to follow DNV 2.7-2 Section 4.11.

Requirements for escape panels have also been modelled on DNV 2.7-2 Section 4.11, capitalising on the experience from the offshore industry.

The Standard covers modules that are either operated by the crew of the host platform, or by teams who may be embarked upon a platform from a different NATO country – thus the requirements for emergency breathing devices and life jackets for these crew have been placed on the platform, where they are best resolved integrally with other escape and evacuation requirements, such as lifeboats.

3.9 HEATING, VENTILATION AND AIR CONDITIONING REQUIREMENTS

Some modules may require Heating, Ventilation and Air Conditioning (HVAC) (e.g. for refrigeration, habitat maintenance, removal of fumes, maintenance of temperature conditions). To simplify platform interfaces, the Standard requires this to be provided as a stand-alone capability on the module, rather than to interface with the platform, for example for provision of chilled water to a module-mounted air treatment unit). However, this may result in higher wild heat within the mission bay, which must be designed for – thus the module information package shall specify the wild heat released. It is common for HVAC equipment to be deployed outside the footprint of an ISO container (see Figure 3), particularly for manned modules – if a platform is likely to embark these, then this must be considered.

Figure 3 – Deployed HVAC unit © Marshall Aerospace and Defence Group

3.10 SYSTEM INTERFACES AND CONNECTIONS REQUIREMENTS

Embarked naval modules will offer enhanced capabilities to their host platforms – the extent to which these can be utilised will depend heavily on integrated module and platform design, particularly with regard to module interfaces. The Standard does not attempt to define requirements to this end, but instead offers basic enablers; common locations, shared standards and common requirements for interfaces and connections. The Standard aims to rationalise and standardise the work explored in the development of ANEP-91 [4]. A schematic of the interfaces covered in the Standard is shown in Figure 4 (see end).

A standardised locational framework for the connections is defined as follows; the connections shall be located in an area extending from each "right hand" corner 1 m along either the long or short walls as shown in Figure 5; the connections are to be at an accessible height; and the connections are not to protrude from the module footprint (i.e. recessed).

Beyond this, no locational requirements are specified – this is to take advantage of the inherent flexibility of interface connections (e.g. cables, hoses etc.), though this may result in some minor impact to operations on-board from trailing services. The overall location was chosen as it reflects the general layout of many commercial designs, affords some flexibility to module designers, and allows access to services without working at height (e.g. on the module roof). The Standard is in contrast to other modularity systems where the connection locations for each interface are significantly more prescriptive (e.g. StanFlex).

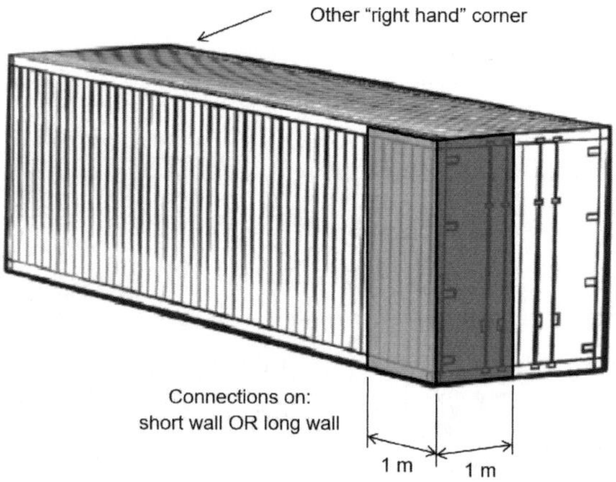

Figure 5 – Interface connections area

Communications and combat systems are likely to present the greatest challenge for integrated designs because of the range of different standards and Combat Management Systems in use across NATO. ANEP-91 (Maritime Mission Modularity) Annex D.19 provides an evaluation of the range of industry standards, their performance, common trends and interoperability; and concludes that industry standard Gigabit Ethernet (GbE) could support the majority of communication requirements for naval modules, and that this would aid rationalisation of the number of individual physical interfaces. The Standard adopts these recommendations, and so GbE is defined as the common interface, over ruggedised RJ-45 connectors. However, it must be acknowledged that adoption of the Standard will be incremental and thus module designers should allow flexibility in interface panels for exchanging connector types, and there may be a need for conversion units for legacy interfaces.

Electrical systems are to be to International Electrotechnical Commission (IEC) standards. These are not naval standards, but were chosen because they are widely used internationally and have precedence in other naval applications. Breakers are to be accessible from the exterior to enable ease of fault-finding, and isolation in the event of an emergency. The issues of electromagnetic compatibility (EMC) and quality of power supply (QPS) are particularly important for naval modules – both for emissions and susceptibility. For example, some modules may require minimum standards of power supply from the host platform, while in other cases platforms may not tolerate emissions arising from embarked equipment e.g. generators or compressors. The definition of requirements for this aspect is an area for further work.

Mechanical services interfaces were considered to be at a lower risk of constraining interoperability, as modules are unlikely to place a high demand on platforms, and pipe fittings can be adapted readily. For this reason, more specific requirements were placed upon mechanical services; for fresh, chilled or grey water these are to be British Standard Pipe (BSP) (to ISO 228-1:2000) or to the "tank-tainer" standards (to ISO 9669:1990); for black water systems these are to be NATO, quick-disconnect, cam-and–groove couplings (to MIL-C-27487D). Practical requirements are defined, such as a minimum size of 1" to prevent accidental damage during handling, appropriate pipe insulation, and the need for external drain ports.

3.11 INFORMATION PACKAGE AND MARKINGS REQUIREMENTS

Naval modules require provision of appropriate information and marking to ensure that they are correctly handled, embarked and connected to host-platform services. The Standard lays out requirements for markings, a standardised information plate and a standardised information package.

Information which needs to be immediately available to an operator is defined for marking onto the exterior of the module. This includes local markings, for example the WLL of any lifting points, inlet / outlet connections etc. but also the requirement for a standardised information plate (example shown in Figure 6 – see end). This plate is defined to show information for identification and for safety – for example fire-fighting, lifting etc.

All other relevant information is defined for inclusion into a module information package, to be kept with the module. This includes layout drawings, equipment specifications, operational constraints and requirements, interfaces, maintenance requirements etc. The module information package is to be used by the Naval Administration to assess compliance prior to authorising embarkation of the module on their platforms.

4. GUIDANCE FOR HOST PLATFORMS AND MISSION BAYS

The Standard is primarily aimed at the requirements for the naval modules themselves – however, authorship of such requirements sometimes led to back-driving requirements for the host vessel, assumptions on the provision of the host vessel, and identification of important information and guidance which needed to be captured. Thus, the Standard includes a section on guidance for host platforms and their mission bays (noting that the scope includes any and all areas on board naval vessels where modules are intended to be embarked – which could include weather decks, vehicle decks, mission bays etc. – but that 'mission bay' is the term employed for brevity). The section includes guidance information and notes which are supplementary to the requirements of the Standard, and are not mandatory requirements. This section of the paper explores this guidance.

4.1 SHOCK

The 'pumpkin mount' shock system is being codified in an additional NATO standard. This system utilises the twist-lock fittings for mounting to the module, and also to the host platform, as shown in Figure 7. Host platform designers should consider the following:

- The module sits in a raised position when mounted on the system – this must be considered for deckhead height and lifting equipment design.
- In a shock event, significant loads will be transferred to the foundation structure.
- The displacement of the module during a shock event must be considered for service connections, inter-module spacing, and overall mission bay layout designs (minimum spacings are recommended).

Figure 7 – 'Pumpkin mount' system © Thornton Tomasetti

4.2 SPATIAL CONSIDERATIONS

There is a prevalence of naval modules with fold-out, slide-out, or expanding/extendable sections which protrude from the envelope of an ISO container when deployed (e.g. HVAC equipment or off-board system deployment mechanisms – see Figures 8 and 9). Where such systems are envisaged for use, mission bays must take account of the attendant space requirements.

The mission bay crane / mechanical handling system will also need to be designed with a height of lift suitable to accommodate the fixed lifting arrangements within the Standard. Where this cannot be achieved, or mission bays are already existing, then alternative arrangements may be required, for example standard ISO container twist-lock lifting accessories, jack-up wheels, or skids.

Figure 8 – Deployed accommodation / office module unit © Marshall Aerospace and Defence Group

Figure 9 – HUGIN off-board system being deployed from module on Korean Naval vessel © Kongsberg, 2010

4.3 INTERFACE CONSIDERATIONS

The services and interfaces for mission bays should be designed to take account of the modules intended to be embarked, particularly where multiple connections are required. The wild heat arising from modules should be understood and factored into the HVAC design for the mission bay, particularly where modules with embarked HVAC are anticipated.

5. NEXT STEPS

The Standard has been reviewed by Royal Navy service personnel, UK Naval Authority and NATO Ship Design Capability Group members, with comments collated for consideration. The overseas comments are particularly welcome as they help ensure international suitability.

The next logical step is to conduct a test-case design of a naval module against the Standard, with input from specialist manufacturers, to examine the cost implications and design interfaces arising from the requirements. This will likely be UK-centric and examine Royal Navy requirements for future modularity. The learning from this exercise should be considered and recommendations made for update of the Standard. Further engagement with the off-shore industry to learn from their experience of employing DNV 2.7-2 would also be highly beneficial, particularly where any limitations of this standard have been highlighted.

Further work on shock / blast loading has been recommended, particularly practical shock testing of an offshore container in accordance with DNV 2.7-2, to better understand the survivability of a naval module – the UK Naval Authority is organising these next steps. The NDP also plan to undertake further work to update the air transport requirements, and aim to scope the development of suitable EMC / QPS requirements for inclusion into the Standard.

The final draft is intended to be released in 2019 to undergo the formal NATO standard approval route.

6. CONCLUSIONS

This paper has identified the key requirements for naval containerised modules as included in the Standard authored by NDP. The origins and rationale for these requirements have been explored, and the need for further work to define these standards has been identified – particularly for shock / blast loading, EMC / QPS, and transportation by air. The key guidance for host platforms, arising from the module requirements, has also been discussed.

7. ACKNOWLEDGEMENTS

The authors wish to acknowledge the role of David Manley of the UK Naval Authority in the NDP's work on modularisation, and his efforts and assistance towards this paper.

8. REFERENCES

1. MANLEY, D., 'The NATO Drive to Mission Modularity', *Proceedings of RINA Warship 2018 Conference*, 2018.

2. 'ANEP-77 Naval Ship Code', *NATO Standardisation Office*, 2019.

3. 'Standard for Certification No. 2.7-2 Offshore Service Modules', *Det Norske Veritas*, 2013.

4. 'ANEP-91 Standard Interfaces for Mission Modules, *NATO Standardisation Office*, 2016.

9. AUTHORS BIOGRAPHIES

Hayden Cole (CEng MIMechE) is a marine engineer for the NDP, seconded from Babcock International where he is a senior design engineer. He is delivering pre-concept designs for future Royal Navy vessels while supporting modularisation tasks. Previously, he completed the first release of the Standard herein.

Julia Alvarez-Mari (CEng MIMarEst ASEP MINCOSE) is a Principal Engineer for the NDP, and leads the Marine Systems team. She has carried out a range of modularity-related studies as part of this team, and was responsible for the initial scoping work for the Standard herein.

Categorisation	Definition	Minimum fire detection system requirement	Minimum firefighting system requirement
No fire risk	Modules containing no energy sources (i.e. electrical, chemical or mechanical)	None	Portable extinguishers
Low fire risk	Modules for storage of non-flammable equipment such as tools, mechanical parts and electrical equipment not connected to power. The module itself contains only a minimum of necessary electrical equipment for lighting, heating, etc. A fire is not likely to be supported.	Fire alarm, non-integrated	Portable extinguishers
Medium fire risk	Modules containing electrical panels, electrical testing equipment, offices. Routinely manned modules (not accommodation).	Fire alarm, integrated	Portable extinguishers. The need for a fixed firefighting system is dependent on risk assessment conclusions
High fire risk	Module containing Category-A machinery spaces, accommodation and stores for flammable liquids. Also high voltage electrical machinery including battery charging and storage.	Fire alarm, integrated)	Portable extinguishers Fixed firefighting system

Figure 2 – Fire risk categorisation table

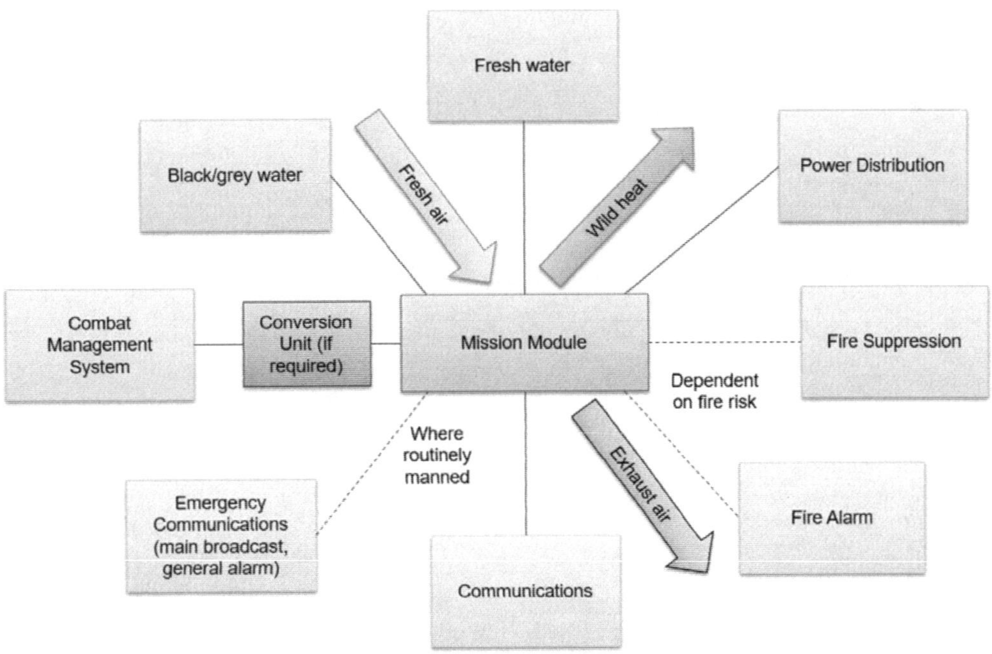

Figure 4 – Module interfaces defined in the Standard

NAVAL CONTAINERISED MODULE
Name of Manufacturer:
Manufacturer Contact Details:
Originating Naval Administration:
Design Assessment Ref:
Date of Assessment:
Serial No:
Compartment Cat/Ship Cat (ANEP-77):
Fire Extinguishing Medium: *Type/On-module/Assumed Platform Provided*
Hazardous Area Rating:
Operational Temperature Range:
Max No of Personnel:
Fitted out Mass:
Centre of Gravity:
WLL of installed lifting equipment:
Information Package Ref. No:
INTERFACE INFORMATION IN MODULE INFORMATION PACKAGE

Figure 6 – Standardised Module Information Plate

STUDIOUS SHIP DESIGN: A REVIEW OF UCL MSC WARSHIP DESIGNS

R J Pawling, UCL, UK

SUMMARY

It is more than 50 years since the UK MoD teaching of warship design moved from Greenwich to UCL, and the course has evolved into MSc Naval Architecture and MSc Marine Engineering courses covering both warships and other complex service vessels, with students from navies, governments and industry worldwide. This presentation will describe this years' designs, outlining the technical solutions proposed to a diverse set of user requirements set by the academic staff. Some of the educational aspects and challenges of the design exercise course will also be described.

Portions of this paper are reproduced from reference [1]; which provides a more detailed example of the work put into a typical MSc SDX ship design.

1. INTRODUCTION: SHIP DESIGN TEACHING AT UCL

Ship design teaching at UCL is provided by the Naval Architecture and Marine Engineering (NAME) group in the Department of Mechanical Engineering. Historically, undergraduate courses were offered to 3 year (BEng) and 4 year (MEng) level, and postgraduate MSc. After a 7-year hiatus in undergraduate teaching of the subject, it has returned in the form of the "Maritime Design" component of the modular Integrated Engineering Programme. The undergraduate ship design as always revolved around a design exercise for a constrained "type" ship, such as a light frigate or container ship, whilst the MSc course is far broader in the range of ship types considered. Student groups have on occasion described their designs in papers to conferences [4], and a multipurpose coast guard vessel was used as an example of the importance of decision making in ship design education by the first author [1].

2. THE MSC SHIP DESIGN EXERCISE (SDX)

The 12-month full-time MSc Naval Architecture and MSc Marine Engineering courses at UCL three main elements; six months of academic teaching and exams; the group Ship Design eXercise (SDX); and an individual project [2], [3]. The SDX is a 45-credit module running for three months full-time between April and June, requiring approximately 450 hours of individual effort, allocated roughly 25% to lectures and visits and 75% to project work. Students work in groups of 2-4 to develop a concept ship design. The small size of the groups helps ensure that all students have visibility over all parts of the ship design process, to avoid a student specializing in only a single aspect of the ship design.

Each group has a different set of design requirements, mostly for warships and service vessels (due to the students' background), and the requirements are characterized by being challenging and relatively open. The students are provided with a User Requirements Document (URD) developed by the teaching staff, in consultation with industry. The URD sets very high-level requirements which can be met in a variety of ways. Few numerical parameters or equipment items are specified, and there are usually several "special studies" or constraints, which may include technologies not yet in operational use. The 2018-19 year saw the introduction of a "Background and Context" document, which describes the general world in which the student designs will be operating; e.g. very strict emissions requirements that lead to some fuels becoming unavailable.

A general overview of the main stages in the MSc SDX is shown in Figure 1, including how they relate to typical stages in a real ship design process. A feature of the SDX is the use of several design reviews and consultancy days with academic staff and external subject matter experts. The students must present the progress of their design and answer questions, with a particular emphasis being on the development of their ability to make design decisions and justify them, rather than the precise technical nature of the solution. The UCL MSc SDX groups consist of both naval architects and marine engineers, and each design review covers both domains, and the interactions between them. The whole-ship design decision making is expected to involve inputs from both domains, with more detailed technical analysis being specific to each MSc.

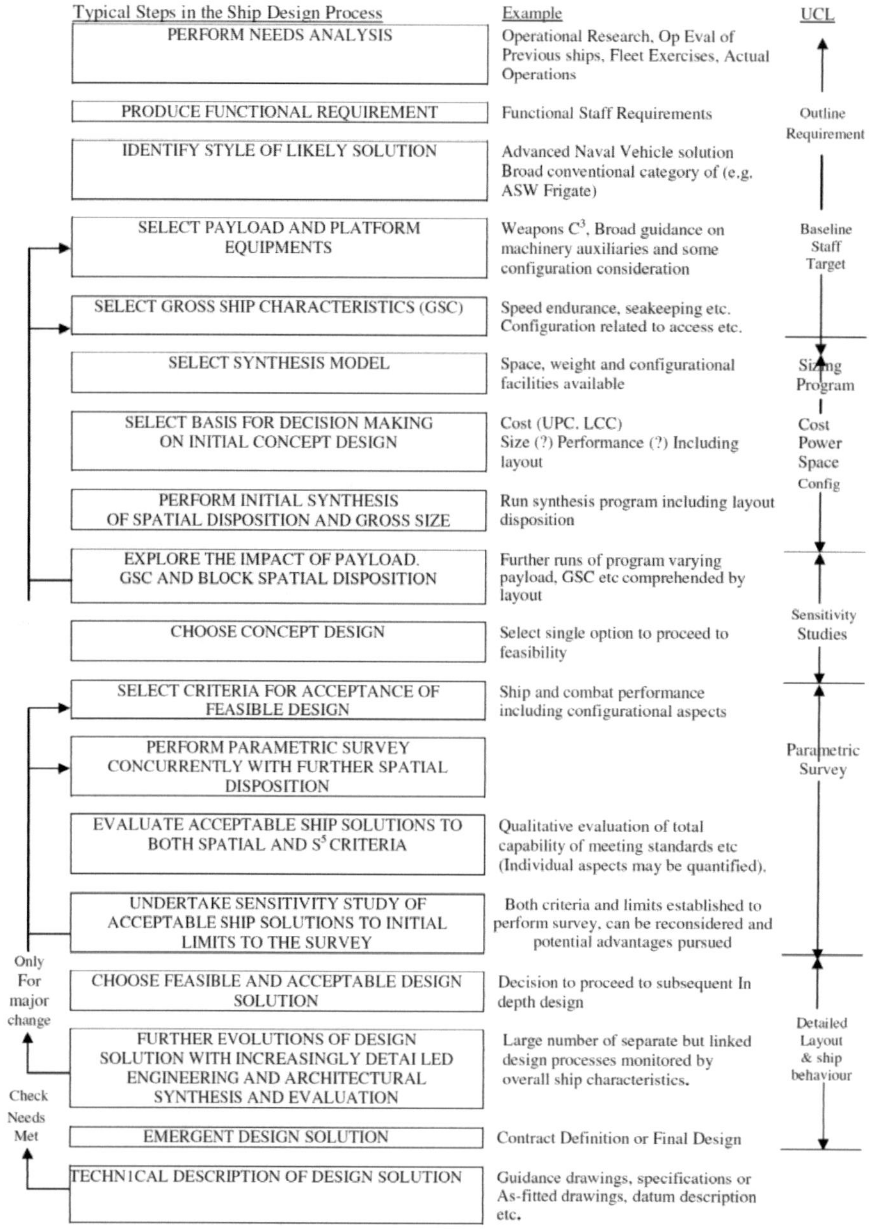

Figure 1: An overview of the MSc SDX process compared with a generic ship design process [5]

2. 2018-19 DESIGN REQUIREMENTS

Table 1 summarises the seven design requirements issues for 2018-19, and the makeup of the groups. Three designs were purely military, with two being a combination of military and civil/policing, and the final two being completely civilian. Table 2 describes the budget allocations for each design requirement. Costs are calculated using UCL data, which is largely simple weight-based costing, but also includes aspects such as shipyard learning and so can be used to evaluate class costs. Historically there has been some uncertainty as to the accuracy of UCL cost estimates, and this was accepted within the context of the SDX as a teaching tool, but it should be noted that the budgets are usually taken from comparison with existing or recent vessels, and baselining exercises are carried out to improve accuracy when additional data becomes available.

Group	Design Requirement	Outline	Naval / Civil	Group Size
1	UxV Carrier	A mothership for multiple types of un-crewed vehicles (UxVs) with particular emphasis on systems consisting of combinations of aerial, surface and subsurface vehicles and through life flexibility and reconfigurability	Naval	2 NA 1 ME
2	Future Missile Fighter (Heavy)	A vessel to provide local and area air defence against a range of projected high-speed above-water threats	Naval	2 NA 1 ME
3	Future Missile Fighter (Light)	A high speed (40-50 knots), advanced technology minimally-crewed surface combatant employing an "aircraft-type" swing-role approach where the vessel can be reconfigured to provide a single high-end capability at a time	Naval	2 NA 1 ME
4	Disaster Response Vessel	An emergency response vessel to be operated by a national government or NGO, providing a range of aid and response to natural and anthropogenic disasters	Both	2 NA 2 ME
5	EEZ Presence Vessel	A multi-purpose patrol vessel for control of a nations' EEZ with some degree of modularity and very low emissions	Both	2 NA 1 ME
6	Short Sea Shipping Ro-Ro	A roll-on-roll-off (Ro-Ro) short sea shipping vessel using automated guided vehicles and having ultra-low emissions	Civil	2 NA 1 ME
7	Future Live Fish Carrier	A vessel to transport live fish from future offshore fish farms, with some design adaptability	Civil	2 NA 2 ME

Table 1: Summary of design requirements for the 2018-19 UCL Ship Design Exercise

Group	Ship	Budget, Millions of Pounds (M£)	Number of Ships
1	UxV Carrier	600 M£ for class at build	3 or 4
2	Future Missile Fighter (Heavy)	9000 M£ for class	8 minimum
3	Future Missile Fighter (Light)	3000 M£ for class	Target of 20
4	Disaster Response Vessel	450 M£ for class of ships plus 50 M£ for a set of naval-specific modules	4
5	EEZ Presence Vessel	225 M£ for class	3
6	Short Sea Shipping Ro-Ro	450 M£ for class	3 is preferred option
7	Future Live Fish Carrier	60 M£ for class	2 to 5

Table 2: Budget allocations for the 2018-19 UCL ship design requirements

3. EXAMPLE DESIGNS

At the time of writing, the 2018-2019 SDX is not complete, so this short paper will only provide high-level information on some of the designs.

3.1 FUTURE MISSILE FIGHTER – HEAVY

This requirement has been dominated by an above-water threat including an attack by multiple, simultaneous hypersonic anti-ship missiles. This has led to an 11,000 tonne solution featuring multi-function radars with a 45m height above the waterline, and an outfit of 130 VLS cells to ensure sufficient Probability of kill (Pk) against the air threat. To assess the performance of the design a UCL-developed engagement modelling tool, "MEM1D" [5] has been used. The uncertainty as to future fuel types, and possibility of changes through life, led the marine engineer to consider a wide range of options, including nuclear and conventional steam, as boilers can be configured to burn a very wide range of fuel types. Figure 2 shows the design at an early stage, clearly illustrating the impact of radar height requirements.

Figure 2: FMF-H CAD rendering

3.2 FUTURE MISSILE FIGHTER – LIGHT

Whereas the FMF-H solution rapidly become dominated by a single aspect of performance, FMF-L has been a more complex decision-making process. The high speed requirement initially appeared to be driving the design, but the student team were able to demonstrate that it could be met, but only by relaxing the initial size constraint of 2500 tonnes deep displacement.

Figure 2 shows the overall geometry of the FMF-L, which uses a trimaran configuration to reach approximately 45 knots with waterjet propulsion. Although superficially similar to the US Navy Littoral Combat Ship (LCS), FMF-L is oriented towards a high-end warfighting capability with strike length VLS, rather than a large mission bay.

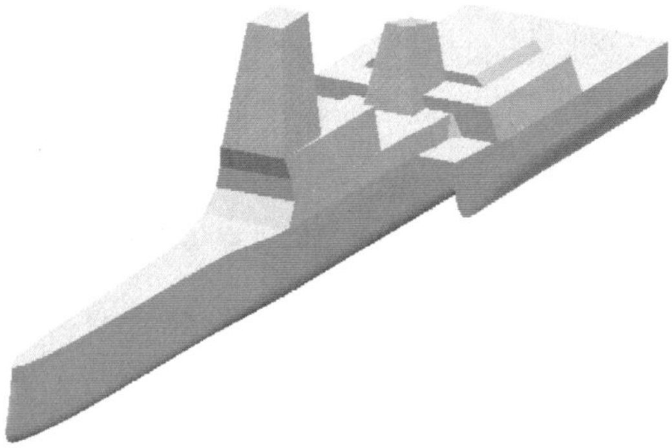

Figure 3: FMF-L overall geometry

3.3 DISASTER RESPONSE VESSEL

Disaster relief ships have been examined by UCL students in previous years, but the 2018-19 requirement was for a vessel designed first for disaster recovery, then with the second role to be determined by the student group. Further, an additional fixed budget was available for modularised military systems, which could be distributed according to the students' cost-benefit analysis. Figure 3 shows the general configuration of the DRV, which was designed to act as a cargo Ro-Ro in everyday service.

Figure 4: Disaster Response Vessel

Although largely civil in design, the DRV features specific areas of complexity in system design, reflecting the requirements of the emergency and military roles. The naval and civil naval architects were encouraged to think about the design requirements of the three contrasting roles and determine which should and should not be included in the basic design. Figure 5 illustrates the use of the Ro-Ro deck for the hospital ship operational scenario.

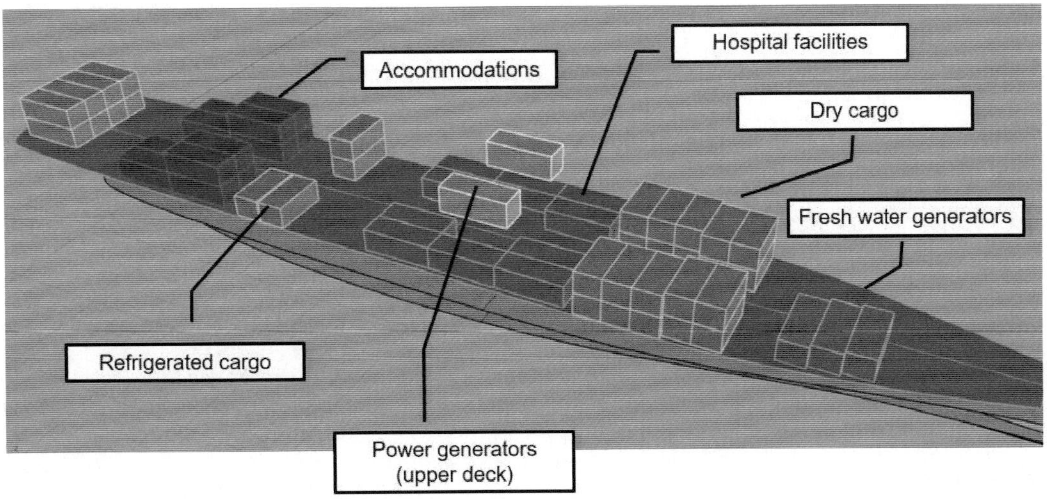

Figure 5: DRV ro-ro deck configured for hospital ship operations

4. ACKNOWLEDGEMENTS

The author would like to acknowledge the UCL MSc students who worked so hard on their ship designs for the last 3 months, providing the source material for this paper.

5. REFERENCES

[1] Pawling, R.J., Bilde, R., Hunt, J., "HYDRA – multipurpose ship designs in engineering and education", International Marine Design Conference (IMDC) 2018, Helsinki, Finland, June 2018.

[2] "MECH0069 Ship Design Exercise for Naval Architects", https://mecheng.ucl.ac.uk/study/modules/mechgn05-project-management-design/

[3] "MECH0084 Ship Design Exercise for Marine Engineers", https://mecheng.ucl.ac.uk/study/modules/mechgr91-ship-design-exercise/

[5] Andrews, D.J. 1986. An Integrated Approach to Ship Synthesis. Trans. RINA. Vol.128, London: RINA.

[4] Collins, L., & Ward, P., "The Affordable Modular Frigate – A Design Study", International Conference Warship 2012: the Affordable Warship, RINA, Bath, June 2012.

[5] Bradbeer, N., 2012: Missile Engagement Model – 1D, Computer code, UCL, London

6. AUTHORS BIOGRAPHY

Dr. Rachel Pawling is a Lecturer in Ship Design at UCL, teaching the subject of ship design to undergraduate and postgraduate students. She obtained her PhD in computer aided ship design in 2007 and has subsequently continued her research in projects funded by the EC, UK and US governments.

MISSIONEASE BOAT HANDLING SYSTEM FOR MULTI-ROLE MISSION SHIPS

R A Wigand and A Kalve, Vestdavit AS, Norway

SUMMARY

Modern mission ships need the flexibility to handle multiple boats, special loads and containers but most multi-boat handling solutions today involve overhead cranes on deck and swinging loads which can be influenced by ship motions and braking; slinging and unslinging boats can be both hazardous and inefficient.
A sheltered 'hangar' located within the vessel's freeboard and accessible from both sides allows boats to be stored, prepared and launched in a safe and dry environment. Vestdavit has devised MissionEase as a unique storage and handling system to transfer boats and other equipment using a patented, deck-mounted multi-boat cradle system that feeds boats between the storage area and davits.
This paper explores how, MissionEase avoids the risks to crew associated with swinging loads, leads to a lower vessel centre of gravity and at the same time achieving a more time-efficient solution for multiple boat handling across the widest array of ancillary craft.

KEY WORDS

davit; boat handling; self-tension; shock absorber; launch and recovery; LARS; safety; FRC/MOB

1. INTRODUCTION

A growing number of modern specialised ships - including those deployed for naval, coastguard, humanitarian, offshore, seismic, and expedition cruise duties - need the flexibility to handle multiple boats, special loads and containers. Increasingly, ship designs are incorporating the mission bay – a sheltered hangar located within the vessel's freeboard that is accessible from both sides the allows boats to be stored, prepared and launched in a safe and dry environment.

Vestdavit has developed the MissionEase, a unique storage and handling system working within the mission bay to transfer boats and other equipment between the storage area and the davits used for launch/recovery. Rather than using overhead gantry cranes, MissionEase uses a system of hydraulic cradles and wagons to move boats from stowage positions to preparation, or launch areas, linking seamlessly with davits to deploy/recover boats from either side of the vessel.

This paper will show how, MissionEase avoids the risks to crew associated with swinging loads. It also presents MissionEase as a more time-efficient solution for multiple boat handling across the widest array of ancillary craft, including RIBS, ROVs and Unmanned Support Vessels.

2. MISSION SHIP DEMANDS

Humanitarian disaster, piracy, terrorism, mines, oil spills, research and dive support - just some of the areas where 'daughter' craft, now play a routine role in the offshore, naval and security sectors.

It is fair to say that the safety and efficiency with which these 'mission ships' launch and recover multiple vessels, whether they be patrol boats, leisure craft, or AUVs and USVs, is a defining feature of their performance overall safety and efficiency. For example, based on its ability to retrieve migrants from the water, a single offshore supply vessel committed by Norway to Operation Triton in the Mediterranean rescued 34,000 people during a two-year tour of duty.

Most recently, as a new generation of smaller cruise ships and yachts designed for polar waters has entered the market, the ability to launch and recover RIBS and other excursion craft has also become part of the expedition experience. Mostly, passengers step onto these 'daughter' craft from ship staircases but getting these vessels onto the water in the first place is another matter. A typical solution from Vestdavit is to be found on the Lindblad expedition vessel Quest, where Vestdavit TSB-2500 davits deploy and recover Zodiacs astern.

A truly multi-role mission bay will enable ships to swap roles quickly and to store and deploy a range of assets on either side of the vessel safely and quickly. One unifying characteristic that owners of these very different vessel types consider including is the 'mission bay' – a sheltered hatchway above the ship's waterline that extends across full beam and is accessible via both sides of the vessel. With davits mounted at the ship's side to lower and lift boats to/from the sea surface, the mission bay provides a protected area to store, maintain and prepare them for launch and receive them after recovery.

Where a ship needs to launch and retrieve a single boat (for example, a man overboard boat), storage can be accommodated within the frame of the davit. However, where multiple boat types need launching and no mission bay is present, boats must be stored on deck and the conventional means of transferring them to/from the davit uses deck cranes (Fig. 1) or overhead gantry cranes.

Figure 1:Typical deck-mounted single point davit/derrick crane

Both solutions load up the vessel's superstructure, with the raised centre of gravity having a negative impact on vessel stability. They also involve suspended, swinging loads on deck that pose a risk to personnel and property.

Furthermore, these solutions take up valuable deck space, and launching boats takes time as handling to/from the davit is time consuming.

Until now, most multi-boat handling solutions have involved use of overhead gantry cranes. Such systems involve swinging loads, which can be dangerous as loads are influenced by ship's motion and braking.

Again, with around 70% of marine accidents linked to human error according to IMO, it is worth considering that slinging and unslinging boats can be hazardous, as well as involving a time penalty.

In addition, installation of overhead gantry cranes would require the reinforcement of the upper section of the mission bay and reducing deck height.

Figure 2: Shipboard gantry crane. If installed in a mission bay, the ceiling structure requires reinforcement, while the risk of sway remains

3. SAFE MISSION BAY HANDLING

Vestdavit has developed MissionEase to offer an integrated handling solution to transfer multiple RIBs, ROVs, Unmanned Support Vessels or similar between the storage and launching/retrieval areas within a mission bay of standard deck height of as low as 2,900 mm at sea.

The system enhances both the safety and the efficiency of operations by turning the transfer operation of boats between storage and davit on its head, to run on floor-mounted rails, rather than ceiling-mounted gantries.

MissionEase consists of a multi-boat cradle system located in the deck of a vessel's mission bay that brings the boats forward from their stored position at the rear of the bay to a transverse rail connecting to the preparation or launch areas (Fig. 3).

Figure 3: MissionEase, basic configuration

Using MissionEase, a single system can feed davits either side of the mission bay to deploy up to three boats within 10 minutes, where a conventional gantry configuration would manage only one deployment.

From its conception, MissionEase has been designed for the widest array of ancillary craft, including RIBS, ROVs and Unmanned Surface Vessels (USV) Autonomous Underwater Vessels (AUV), accommodating a range of hull profiles, but also underwater vehicles featuring stabilisers or even frames.

From the safety perspective, the rails for MissionEase are significantly lighter than an overhead gantry assembly (Fig. 4); With its own braking system, MissionEase allows boats to be moved within the bay, even in high seas or with a list on the vessel for launching, or back to stow when recovered.

This is safer, and brings a cost saving as one can handle several boats with the same davit system.

Figure 4: The MissionEase transfer mechanism is in the floor of the mission bay, reducing risk, the need for structural reinforcement, and cost

MissionEase allows boats to be moved safely and quickly on cradles within the mission bay, even in high seas or during a vessel listing. This has specific benefits:

- for deck crew; avoiding swing risks associated with overhead cranes used for the deck transfer of mission boats.
- for the 'human in the loop'; semi-automation of the transfer process to the 'ready-to-launch' position avoids human error (Fig. 5).
- for the launch boat crew/passengers; easier and safer boat embarkation from sheltered area - quicker return to protected area in heavy seas/hostile environments (Fig. 6).

In addition, a single operator can control the entire MissionEase handling cycle using a Remote-Control unit, either hardwired or radio remote – from storage to engagement with the davit, and on to set down on the water, or retrieval back to stow when recovered. All boat movements during the transfer process are aligned with the movements of the ship itself, even in the roughest seas. The Telescopic Davits that are integral to the system, are all delivered with shock absorbers and guiding arms for safe handling of boats and crew. The system also benefits from a manual back-up, with accumulators built-on for emergency operations in case of blackouts.

Figure 5: MissionEase enables a high degree of automation transferring to/from boat launch/retrieval

4. DETAIL AND FIRST APPLICATION

At the heart of MissionEase is its patented feeding system, which is powered either electrically or using hydraulics and links seamlessly with either dual or single-point davits - selected based on the vessel's boat handling requirements. capacity.

The feeding system holds the boats in a series of storage cradles that are moved forward from the stowage on rails, before transfer to a transverse rail which allows them to move starboard / port to link seamlessly with dual or single-point davits on either side of the vessel. The Telescopic Davit Structure is assembled as one complete welded unit with blocks and bearings. It will be attached to the ships ceiling by six welded hinges. Forces on the deck structure depend on the systems SWL.

A MissionEase system will normally be hydraulically operated, powered by a dedicated HPU between 80 and 200 Kw. The Hydraulic Power Unit is of the submerged type, where pump and electrical motor are submerged in a stainless-steel tank. This reduces noise and brings maintenance costs down to a minimum.

The electrical cabinet, with the davit soft starter, is prepared with all necessary electrical connections, and need to be placed inside the ship.

Already patented in Norway and with an application pending to cover all EU territories, Vestdavit has also secured a patent from the United States Patent and Trademark Office for the MissionEase feeding system which protects the multi-boat handling system's unique mode of operation from plagiarism.

The solution is fully type approved by DNV-GL in capacities ranging from 10,000kg to 40,000kg. MissionEase is CE-marked according to the EU's Marine Equipment Directive. Approvals by other classification societies are available on request.

The Telescopic davits used as part of MissionEase are certified as aligning with latest SOLAS requirements.

Figure 6: MissionEase engages seamlessly with the shipside davit

All davits from Vestdavit are supplied with Self-Tension (wave compensation systems), Shock absorber, Anti-Pendulation Guiding Arms, Wire Puller, Hydraulic End Stop, and remote controls.

The overhead telescopic dual point davit can operate both in dual point and single point configuration. It has synchronized movement of the two independent davit arms when operating in dual point configuration, including dual independent operated winches with dynamic brakes. Hoisting and lowering speed is recommended at 40 m/min with variable speed 0-40 m/min.

MissionEase was introduced in 2015, with a first order placed in 2016 for a crew change boat for offshore vessel owner Esvagt, with the boat transfer system built in to an 88.4m length crew change vessel delivered by Astillero Zamakona. The transverse MissionEase cradle will feed boats to one Vestdavit TDB-5000 telescopic davits (Fig. 7) on each side of ship.

Figure 7: Vestdavit TDB-5000 telescopic davit

This system has capacity for four boats. It will be adaptable, accommodating three different boat types, plus one transverse cradle with lift and lowering capability. The cradles move across the ship, picking up boats from the supports and feeding them to the davits for launching.

Multi-boat handling using MissionEase is expected to enhance the attractions of OSVs for charter. When multiple ships are involved in the same mission, MissionEase could enhance fleet responsiveness by allowing individual ships to pick up boats belonging to other vessels rapidly. Capability like this is will enhance the attraction of ships featuring MissionEase in the charter market.

Later deliveries are expected to include longitudinal cradles, with MissionEase able to stow, launch and retrieve from 5 to 15 different RIBS, daughter boats and Unmanned Surface Vehicles safely and quickly in high seas from within a ship's protected hangar. The solution was launched into the luxury passenger vessel market in 2017, where it can stow, launch and retrieve up to 24 RHIBs, kayaks, waterjets, or similar.

More than 20 'adventure' cruise ships or yachts were reckoned to be on order at the end of 2017. Vestdavit's MissionEase multiple-boat handling system for hangars and mission bays adds a new option to the expedition ship mix.

Figure 8: Davit handling at ship's side after transfer to the lifting/lowering position

5. WIDER UPTAKE

In November 2018, Vestdavit secured a major breakthrough for MissionEase, after winning the contract to equip the Nexans flag ship cable layer *Nexans Aurora*. The high-end Nexans cable laying vessel (CLV) has been designed by Skipsteknisk in Ålesund and will be constructed by Norwegian yard Ulstein Verft. At 149.9m in length, 31 m width and with capacity to accommodate 90 personnel, she will perform power cable laying, cable system protection and trenching, and be capable of high manoeuvrability and DP3 station-keeping capabilities.

This complete MissionEase system order is most significant, because this vessel has been designed specifically to operate in rough conditions.

The full-size Vestdavit MissionEase system will be able to handle several workboats/tugboats for launch from or recovery to either side of the ship, with one operator using a remote control to launch and recover the mission boat, including all on-deck handling. The davits to be used in combination with MissionEase are Vestdavit's compact and reliable Telescopic TDB-5000 units, which are often chosen by Coastguards, Navies and offshore operators when space, as well as performance, is key. The TDB-5000 is also equipped with a range of safety features such as shock absorbers, constant tension, remote control and wire puller.

6. CONCLUSIONS

Given the growing requirement for a wider variety of ships to feature the mission bay, Vestdavit believes the development of MissionEase is of significance across the naval, patrol, offshore and passenger sectors. It contends that the conventional solutions used to transfer boats to/from the davit are sub-optimal in terms of safety and efficiency, and that these shortcomings run contrary to the fact that mission bay operations often define the ability of mission ships to perform their duties effectively. In seamlessly feeding boats between the storage area and the dual or single-point davits on either side of the vessel, MissionEase optimises the efficiency and safety of a central ship function.

Vestdavit designs, supplies and supports tailor-made solutions for launching and recovering boats in difficult conditions at sea. Its range of boat handling systems and davits are the first choice of navies, coastguards, seismic survey operators, pilot authorities and offshore operators who need to be able to operate small boats safely from larger vessels. Since 1975 Vestdavit has supplied over 1,900 davits and side and stern launch systems. They have proven themselves for almost 40 years of use in the North Sea and other harsh environments around the world. Self-tensioning and shock absorbing systems ensure crew safety and widen the operational window for the users.

www.vestdavit.no

7. AUTHORS BIOGRAPHY

Rolf Andreas Wigand leads the management and business development at Vestdavit, the Norwegian boat handling specialist whose davits provide safe, efficient and reliable launch and recovery in difficult conditions at sea for navies, coastguards, pilot authorities, expedition ship and offshore operators. His responsibilities include relations with the company's worldwide agency network and Vestdavit's strategy for growth, via new products, new territories, and after sales.

With a degree in Trade Economics from the Oslo Business Academy, and a Master's in Business Administration from Arizona State University, Rolf Andreas has over 35 years of experience in business development and building international relations. His past positions include CEO for the Bergen-based VW dealership, a Marketing Director role at global group LEGO Company SpA, Milan, and posts with Coca-Cola (Norway), Masterfoods (Mars Inc.) and Mitsubishi Motors.

Atle Kalve, Vestdavit Development Director / Naval Architect: Atle Kalve has devoted his working career to developing safer and better boat handling. Kalve holds a degree in Naval Architecture from the Norwegian Technical University. Following several years spent in naval research, Kalve has worked with Vestdavit since 1982, in a variety of commercial roles. Currently, his main tasks relate to product innovation and development, and on ensuring that Vestdavit remains at the forefront of boat handling systems technology. Kalve has also been responsible for a number of inventions which have secured patents, with the MissionEase system being the latest.

INLINE-SLOT EJECTOR DIFFUSER FOR A WARSHIP TO SUPPRESS IR SIGNATURES

L Singh, S N Singh and **S S Sinha,** Indian Institute of Technology Delhi, India

SUMMARY

The infrared (IR) signatures emitted by warships can substantially reduce the survivability chances while operating in the enemy area, as these signatures can be tracked and locked-on by the heat seeking missiles. The primary source for the IR signatures in a warship is the gases emanating from the exhaust of the gas turbine engine. These signatures can be suppressed by installing the passive infrared suppressors such as ejector diffusers, downstream of the turbine exhaust. The prime objective of an ejector diffuser is to reduce the exhaust gases temperature with minimal back pressure. In the present investigation, inline-slot conical ejector diffuser is numerically studied. From the open literature it is found that only step-slot ejector diffuser has been explored and no work on the inline-slot ejector diffuser is reported. Preliminary investigations on the inline-slot ejector diffuser show the performance to be better in terms of mass entrainment and static pressure recovery than the step-slot ejector diffuser, which is commonly used with warship power plants. The focus of the present study is to establish the effect of nozzle exit conditions (i) inlet swirl (S), and (ii) inlet turbulence intensity (TI) on the performance of inline-slot ejector diffuser. In the first part, inlet swirl is varied in the range $0 \leq S \leq 0.3$ in step of 0.05, and in the second part TI is varied in the range of $1\% \leq TI \leq 15\%$ in step of 3%. The performance is evaluated in terms of local and cumulative mass entrainment ratios, non-dimensional temperature distribution, and static pressure recovery. For the first part, it is seen that there is 3% drop in cumulative mass entrainment with the increase in swirl number from 0 to 0.3 and this can be attributed to the drop in potential core region. Higher wall temperatures in the mixing tube are observed for all the configurations with swirl cases. Static pressure recovery increases with the increase in the swirl number. For the second part, the effect of turbulence intensity on the performance of inline slot ejector diffuser is carried out. It is seen that the mass entrainment decreases by ~5% when turbulence intensity is increased from 1% to 15%. No significant effect of turbulence intensity is seen on the temperature distribution and pressure recovery.

NOMENCLATURE

Re_{nz}	Nozzle exit Reynolds number
U_{in}	Nozzle inlet axial velocity (m/s)
v_i	Velocity components (m/s)
p	Pressure (Pa)
τ_{ij}	Total stress tensor (m²/s²)
e_0	Total energy (J/s)
Φ	Cumulative mass entrainment ratio
κ	Local mass entrainment ratio
ψ	Normalized temperature distribution
C_p	Coefficient of static pressure
y^+	Non-dimensional wall distance
ρ	Density of air (kg/m³)
μ	Dynamic viscosity (m²/s)
μ_t	Turbulent viscosity (m²/s)
λ	Thermal conductivity (W/mK)
λ_t	Turbulent thermal conductivity(W/mK)
k	Turbulent kinetic energy (m²/s²)
ω	Specific turbulent dissipation rate (s⁻¹)
T	Local temperature (K)
T_0	Ambient temperature (K)
T_g	Gas turbine exit temperature (K)

Superscript

~	Favre-averaged quantity
-	Reynolds-averaged quantity

1. INTRODUCTION

One of the biggest survivability threats to a warship is the locking and targeting by an infrared (IR) guided missile. The lethal nature and high accuracy of the IR guided missiles makes it intractable, and is often referred as 'fire and forget'. To counter the IR guided missile an infrared suppression system (IRSS) is installed on the warship. One of the vital functions of the IRSS is to suppress the heat signatures originating from the ship. The source of the dominant heat signatures produced by a warship are the exhaust gases where the temperature is >700K. At this temperature the IR guided missiles can easily lock and hit the target. So, IRSS aims to lower the gas temperature to values, <450K, thereby avoiding the IR guided missiles. The device used to perform the cooling of the exhaust gases is commonly known as ejector diffuser. It is installed as an exhaust assembly for the gas turbine [1]. The design of the ejector diffuser is such that it entrains cold ambient air through slots on its surface, and allows mixing with the exhaust gases. The mixing of hot exhaust gases with cold ambient air results in the temperature drop of the mixed gases before exiting to the environment. Figure 1 shows the schematic of an ejector diffuser and highlights the main components (i) nozzle, (ii) mixing tube, and (iii) slotted diffuser. The accuracy of IR guided missiles [2] places higher emphasis on the performance of the IRSS device.

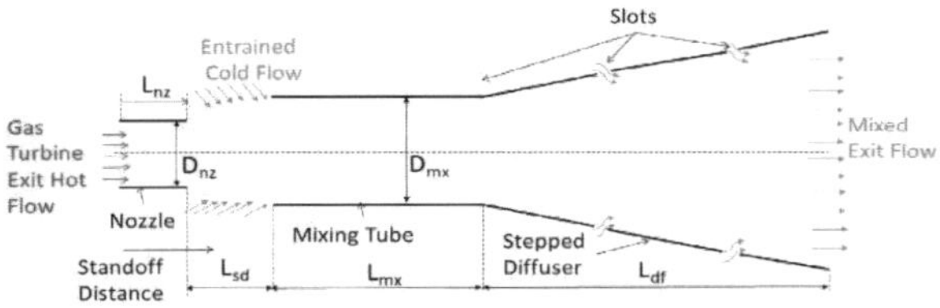

Figure 1. Schematic of a typical ejector diffuser with constant slot width and its components.

Very little information on ejector diffuser is available in open literature due to the confidential nature. However, in the last two decades, some studies on the ejector diffuser have been reported. Most of these studies have investigated the effect of ejector diffuser geometry on its performance. Sen [3] studied a conical stepped ejector diffuser where the effect of various parameters such as length of overlap at step slots, diffuser wall thickness, and inclined slot openings were studied. Singh et al. [4] studied square shaped step ejector diffuser and reported drop in mass entrainment by 26% for wide angle diffuser. In another article [5], effect of number of slots on the performance is reported. The authors have concluded that there is no gain in mass entrainment but observed marginal increase in static pressure recovery. Chen et al. [6] studied oblong shaped ejector diffuser and observed better performance than conical ejector diffuser under moderate swirl conditions. A new shape of ejector diffuser as inline-slot conical ejector diffuser has been studied by Singh et al. [7]. They found better mass entrainment and pressure recovery for these diffusers.

The present study reports the effect of nozzle inlet conditions namely inlet swirl (S) and turbulence intensity (TI) on the performance of inline-slot ejector diffuser. The flow at the gas turbine exit can have weak to moderate swirl (S<0.3). Second part of the study covers the effect of nozzle exit turbulence level on the performance of ejector diffuser.

2. PROBLEM FORMULATION

The present study has been carried out using 'ANSYS FLUENT' code. Table 1 gives the cases investigated. numerical studied. The gas turbine exhaust gases temperature is fixed at 700K and the ambient temperature is fixed at 300K. All the geometric details are kept as constant and are given in Table 2.

Table 1. Cases for the current investigation

Part 1		Part 2	
Cases	Amount of Swirl (S)	Cases	Turbulent Intensity (TI)
S=0	No swirl	TI_1	1%
S=0.05	0.05	TI_3	3%
S=0.1	0.1	TI_6	6%
S=0.15	0.15	TI_9	9%
S=0.2	0.2	TI_12	12%
S=0.25	0.25	TI_15	15%
S=0.3	0.3		

A two-dimensional axisymmetric computational domain is created to conduct the numerical study. Figure 2 shows the computational domain selected for the current study. All

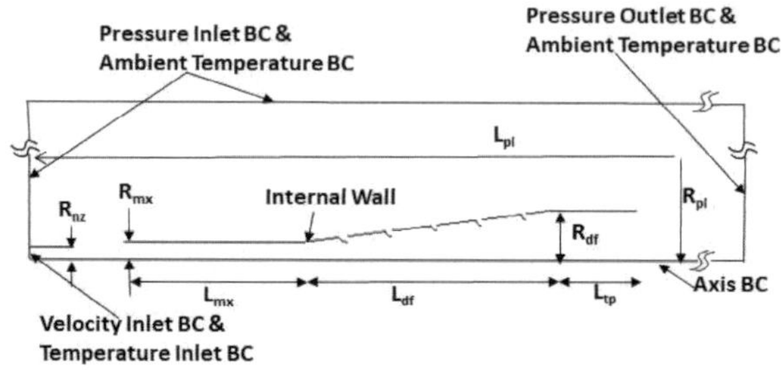

Figure 2. Computational domain and boundary conditions.

the important dimensions are labelled and their values are given in Table 2. Sufficiently large plenum size [6] is created enclosing the ejector diffuser so that plenum boundary conditions has no effect on the flow behaviour in the ejector diffuser. Figure 2 also highlights the boundary conditions applied to all the surfaces. At the inlet swirl or turbulence intensity values are also specified along with other inlet conditions. All other boundary conditions remain same. The computational domain consists of nozzle, mixing tube, inline-slot diffuser, tailpipe, and the plenum surrounding the ejector diffuser. The present study is conducted at a fixed Reynolds number. Re_{nz} is calculated for the flow conditions at 700K (U_{in}=60m/s, ρ_{in}=0.505kg/m3, μ_{in}=3.3889552e-5 kg/m-s). The wall boundary conditions for the ejector diffuser are specified as no-slip for flow variables and thermal coupled wall for the heat transfer variables.

Table 2. Dimension of ejector diffuser

Name	Description	Value
D_{nz}	nozzle exit diameter	50mm
L_{nz}	nozzle length	$1D_{nz}$ [4]
L_{sd}	standoff distance	$2.25D_{nz}$ [8]
D_{mx}	mixing tube diameter	$\sqrt{2.25}D_{nz}$ [8]
L_{mx}	mixing tube length	$8D_{nz}$ [8]
L_{df}	length of diffuser	$11D_{nz}$ [3.4]
L_{tp}	length of tail pipe	$4D_{nz}$ [3, 4]
L_{pl}	length of plenum	$65D_{nz}$ [6]
D_{pl}	diameter of plenum	$21D_{nz}$ [6]
AR_{mx}	area ratio of mixing tube inlet to nozzle exit	2.25[8]

The computational domain is discretized using the structured meshing approach where quadrilateral cells are generated to discretize the computational domain. As the present study employs SST k-ω turbulence model, y+≤1 has been maintained to capture the near wall gradients. The grid quality is ensured through the aspect ratio (AS) of each cell.

3. NUMERICAL DETAILS

This section gives the governing equation, turbulence model selection, validation and mesh independence studies along with numerical details.

3.1 GOVERNING EQUATIONS

The numerical study is carried out by solving the steady state Reynolds averaged Navier-Stokes (RANS) and Favre averaged Navier-Stokes (FANS) equations by using Ansys Fluent. Since the density is not constant it is appropriate to use density weighted averaged values of the variables to absorb any correlation terms of density fluctuation. The set of modelled governing equations for the current study are given from Eq. (1) to (8):

$$\frac{\partial}{\partial x_i}(\bar{\rho}\tilde{v}_i) = 0; \quad (1)$$

$$\frac{\partial}{\partial x_j}(\bar{\rho}\tilde{v}_j\tilde{v}_i) = -\frac{\partial}{\partial x_j}(\bar{p}\delta_{ij}) + \frac{\partial}{\partial x_j}(\tilde{\tau}_{ji}^{tot}); \quad (2)$$

$$\frac{\partial}{\partial x_j}(\bar{\rho}\tilde{e}_0) = \frac{\partial}{\partial x_j}(\tilde{q}_j^{tot}) + \frac{\partial}{\partial x_j}(\tilde{v}_j\bar{p}); \quad (3)$$

$$\tilde{\tau}_{ij}^{tot} = \tilde{\tau}_{ij}^{lam} + \tilde{\tau}_{ij}^{turb}; \quad \tilde{q}_j^{tot} = \tilde{q}_j^{lam} + \tilde{q}_j^{turb} \quad (4)$$

$$\tilde{\tau}_{ij}^{lam} = \mu\left[\left(\frac{\partial \tilde{v}_i}{\partial x_j} + \frac{\partial \tilde{v}_j}{\partial x_i}\right) - \frac{2}{3}\frac{\partial \tilde{v}_k}{\partial x_k}\delta_{ij}\right] \quad (5)$$

$$\tilde{\tau}_{ij}^{turb} = \mu_t\left[\left(\frac{\partial \tilde{v}_i}{\partial x_j} + \frac{\partial \tilde{v}_j}{\partial x_i}\right) - \frac{2}{3}\frac{\partial \tilde{v}_k}{\partial x_k}\delta_{ij}\right] - \frac{2}{3}\bar{\rho}k\delta_{ij} \quad (6)$$

$$\tilde{q}_j^{lam} = \lambda\frac{\partial \tilde{T}}{\partial x_j}; \quad \tilde{q}_j^{turb} = \lambda_t\frac{\partial \tilde{T}}{\partial x_j} \quad (7)$$

$$\bar{p} = \bar{\rho}\frac{R}{\gamma}\tilde{T}; \quad \gamma = \frac{c_p}{c_v} \quad (8)$$

The air properties are defined as function of temperature and are adopted from [9]. The piece-wise polynomial functions for the thermal conductivity (λ), specific heat at constant pressure (cp), and dynamic viscosity (μ), adopted from [9], are shown below:

$$\lambda = 1.00233 \times 10^{-3} + 9.04396 \times 10^{-5}T - 2.90213 \times 10^{-8}T^2 + 4.63995 \times 10^{-12}T^3 - 2.16473 \times 10^{-22}T^4 + 6.96289 \times 10^{-26}T^5; \quad (9)$$

$$c_p = 9.55481 + 6.29068 \times 10^{-1}T - 3.10144 \times 10^{-3}T^2 + 6.92621 \times 10^{-6}T^3 - 6.37568 \times 10^{-9}T^4 + 2.10762 \times 10^{-12}T^5; \quad (10)$$

$$\mu = 2.57183 \times 10^{-7} + 8.54743 \times 10^{-8}T - 1.0303 \times 10^{-10}T^2 + 1.09789 \times 10^{-13}T^3 - 6.76718 \times 10^{-17}T^4 + 1.75005 \times 10^{-20}T^5 \quad (11)$$

3.2 TURBULENCE MODEL SELECTION AND VALIDATION STUDY

Two-equation turbulence modelling is used to address the additional terms after Reynolds averaging. After evaluating different two-equation turbulence models, SST k-ω turbulence model is selected to conduct the numerical study [8]. The complete set of closure equations for SST k-ω can be found in [10]. An additional validation study using SST k-ω turbulence model is conducted by comparing the predicted axial velocity profile with the experimental measurements. The velocity measurement in an ejector diffuser is carried out at multiple axial locations (see Figure 3).

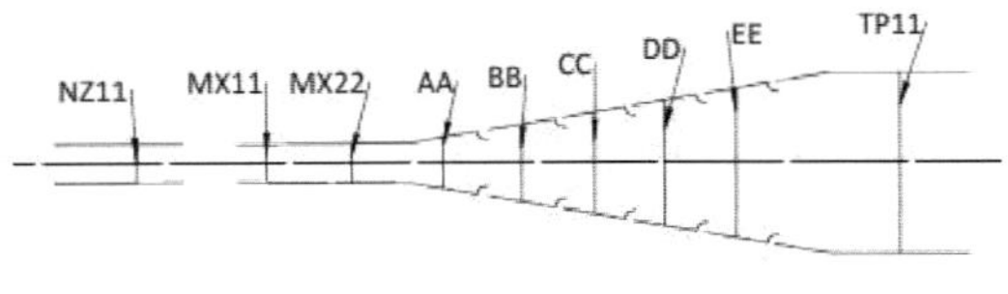

Figure 3. Radial line sections where results are extracted.

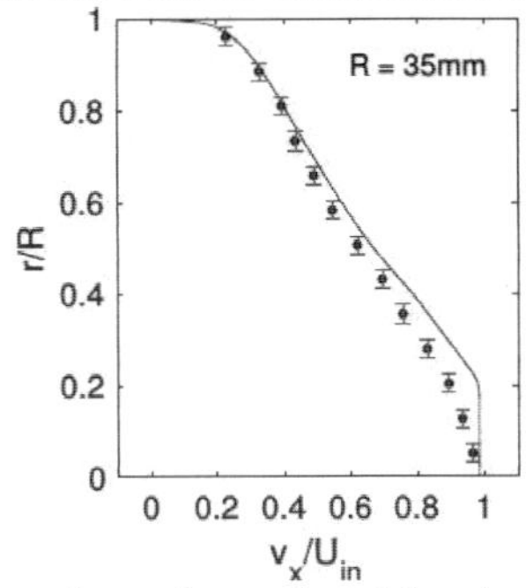

Figure 4. Comparison of experimental and numerical axial velocity profile upstream of first slot.

Figure 4 gives the comparison of axial velocity profile at Location AA which shows reasonable matching. A maximum deviation of 18% local in the axial velocity is observed close to the axis of the ejector diffuser whereas the deviation in the mass-weighted average value of the axial velocity is less than <4%.

3.3 MESH INDEPENDENCE

The mesh independence study and the order of convergence analysis has been conducted for three grids, coarse = 0.5×10^5, medium =1.0×10^5, fine = 1.8×10^5, with the grid refinement factor of 2. Figure 5 shows the

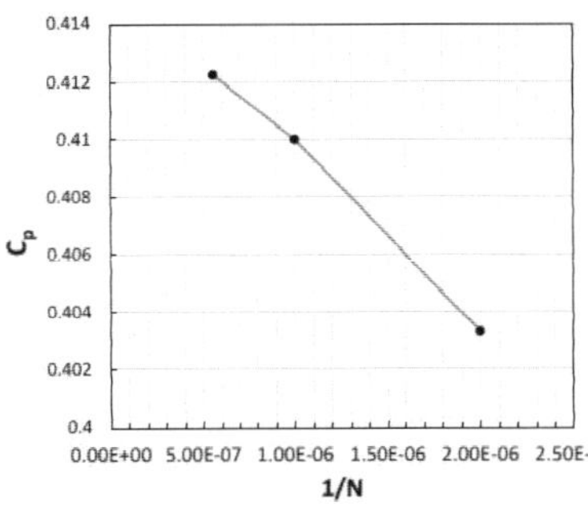

Figure 5. Grid independence and order of convergence analysis as second order scheme.

plot of coefficient of static pressure C_p verses the grid index (GI) factor, where GI=1/N$^{\{2/2\}}$ and N is the total number of cells in a grid [11]. It can be seen that the error between the medium and fine grid is relatively small. Further, the selected order of accuracy (p) for the simulation study is a second-order accurate (p=2) scheme. Figure 5 also shows that C_p for the three grids falls on a straight line, thus depicting the order of convergence as a second order accurate solution. For the current study, medium grid is selected for all the simulations.

4. RESULTS AND DISCUSSION

Results from the numerical study are presented here in terms of mass entrainment, temperature distribution and static pressure recovery.

4.1 EFFECT OF SWIRL

Swirl is introduced at the inlet of the nozzle in terms of tangential velocity, and is quantified as swirl number (S). It is defined as the ratio of the axial flux of angular momentum to the axial flux of axial momentum as defined in Equation 12

$$S = \frac{\int \rho w U_{in} r dA}{\int \rho r U_{in}^2 r dA} \quad (12)$$

where w is the tangential velocity and U_{in} is the axial velocity. Table 1 gives the range of investigation.

4.1(a) Mass Entrainment

The amount of ambient air entrained by an ejector diffuser is an important performance parameter. The mass entrainment in an ejector diffuser depends on the geometrical and dynamical parameters. The amount of mass entrainment is estimated by two ratios (i) local mass entrainment ratio (κ), and (ii) cumulative (global) mass entrainment ratio (φ). These are defined by Equations 13 and 14:

$$\kappa = \frac{m_{je}}{m_{in}}, \quad (13)$$

$$\phi = \frac{\Sigma m_{je}}{m_{in}}, \quad (14)$$

where m_{je} represent the entrained mass flow rate through an individual slot opening and m_{in} is the nozzle exit mass flow rate.

Table 3 presents a comparative analysis of κ for all the cases at each slot. Following observation are made; (i) compared to the no-swirl case (S=0) all swirl cases show marginal improvement in mass entrainment through the standoff distance except S=0.25 and S=0.3, (ii) marginally higher entrainment is observed for all the swirl cases compared to no-swirl case at the first slot, (iii) the mass entrainment for S=0.05 and S=0.1 is marginally higher than S=0 at the second slot, (iv) the mass entrainment for all the swirl cases is lower compared to S=0 from third slot onwards. Further mass entrainment tends to decrease with increase in swirl number.

Table 3: Comparison of local mass entrainment ratio (κ)

	S=0	S=0.05	S=0.1	S=0.15	S=0.2	S=0.25	S=0.3
κ_{SD}	0.974	0.980	0.980	0.980	0.978	0.974	0.968
κ_1	0.151	0.154	0.154	0.154	0.153	0.153	0.152
κ_2	0.293	0.295	0.294	0.291	0.288	0.285	0.280
κ_3	0.390	0.390	0.387	0.382	0.376	0.371	0.364
κ_4	0.522	0.520	0.515	0.512	0.505	0.496	0.487
κ_5	0.879	0.877	0.870	0.868	0.861	0.848	0.837

The driving mechanism for the mass entrainment at any slot depends on the static pressure differential across the opening. The difference in the κ among different cases is understood by observing the static pressure contour plot. Figure 6 shows the comparison of static pressure distribution for S=0 and S=0.3 where difference can be seen at different slots.

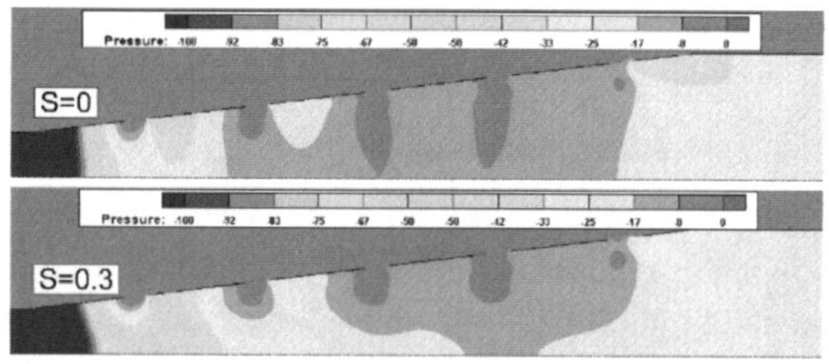

Figure 6: Static pressure contour plot

Figure 7 presents the plot of cumulative mass entrainment ratio as a function of swirl number. It can be seen that with the increase in the swirl number, the value of φ tends to decrease. Except for case S=0.05, φ for all other swirl cases is less than S=0. The maximum drop in the value of φ is observed for the case of maximum inlet swirl S=0.3, where the % difference in φ compared to no swirl case is ~3%. So, it is concluded that any amount of swirl larger than 0.05 will have adverse effect on cumulative mass entrainment. Further, as swirl number increases the mass entrainment decreases.

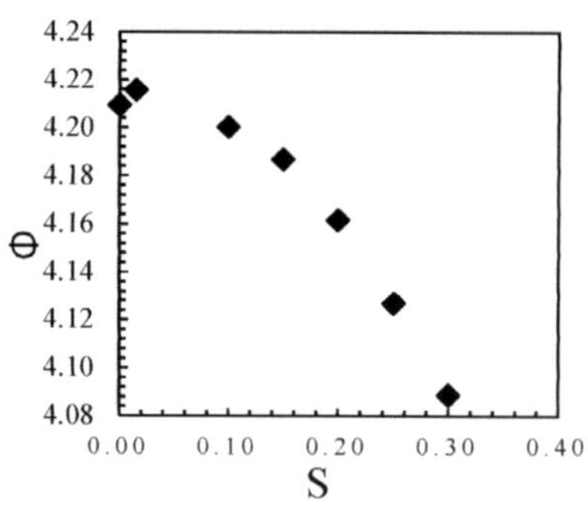

Figure 7: Cumulative mass entrainment ratio (φ) as function of swirl number

4.1(b) Temperature Distribution

To understand the temperature distribution with in the ejector diffuser, a normalized temperature variation (ψ) parameter is studied which reveals the extent of cooling in the ejector diffuser. It is defined as (Equation 15):

$$\psi = \frac{T-T_0}{T_g-T_0}, \quad (15)$$

where T is the local temperature, T_0 is the ambient temperature and T_g is the temperature of exhaust gas at the nozzle exit. The critical areas of interest which contribute to maximum infrared energy are the ejector diffuser exit, core temperature, and the wall temperatures. Figure 8 shows the plot of ψ at the axis of the ejector diffuser for all the cases. The value of ψ=1 represent no cooling while ψ<1 represent lower temperatures than the initial exhaust gas temperature. For all the cases, the value of ψ becomes less than 1 in the range 0.3<z/L<0.4. This represent cooling of the core region. It is observed that the cooling of the core starts early for high swirl number case and it delays with lower swirl number. However, it is observed that all cases after z/L>0.6 show similar value of ψ.

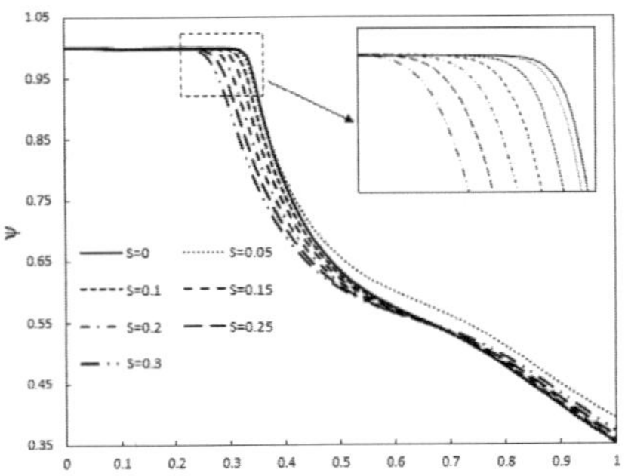

Figure 8: ψ at the core of the ejector diffuser.

Figure 9 presents the variation of ψ at the ejector diffuser exit. For larger part of the cross-section (0.4<r/R<1), all cases show similar profiles of ψ. The difference is observed in the core region 0.4 < r/R ≤ 0 were ψ increases as swirl number increases. However, the temperature corresponding to the maximum value of ψ=0.37 at the core even for S=0.3 is 448K, which is within the safe limits of operation.

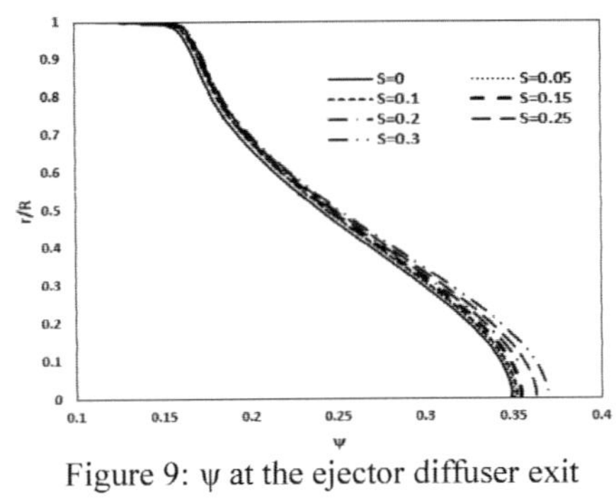

Figure 9: ψ at the ejector diffuser exit

Another critical area of interest is the ejector diffuser wall temperature distribution. Figure 10 shows the variation of ψ along the non-dimensional wall length (z\L). At z/L=0.1, just after the standoff distance, the value of ψ is close to 0. This shows that the wall temperature is approximately equal to ambient temperature (T_0=300K). The value of ψ tends to increase along the mixing tube length (0.1<z/L<0.44), and the profile is different for all the cases. The no swirl case has lower ψ values along the length and it tends to increase with the increase in swirl. This is due to the presence of tangential velocity component in the flow which tends to push the core flow outwards. The maximum value of ψ at the wall is around 0.45. At this value the wall temperature is 480K which is slightly outside the safe limits of operations. However, in the diffuser (0.44<z/L<1) all ψ profiles tends to merge and no significant effect of swirl on the wall temperature is observed.

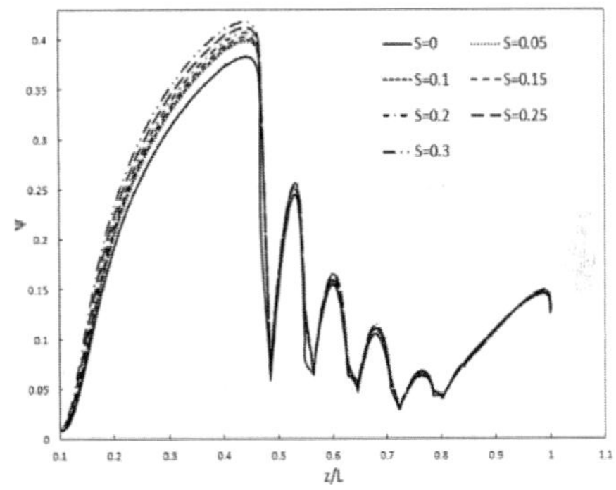

Figure 10: ψ at the ejector diffuser wall

4.1(c) Static Pressure Recovery

Static pressure recovery (C_p) is another important performance parameter for an ejector diffuser as it affects the gas turbine engine performance. High (C_p) is desirable. C_p is defined as the ratio of the difference in static pressure at a given location and a reference location to the dynamic presure at the reference location. In the case of mass entraining ejector diffusers, C_p is modified as suggested in [12]:

$$C_p = \frac{(p_{ex}-p_{in})(m_{in}+\Sigma m_{je})}{\frac{1}{2}(\rho_{in}U_{in}^2 m_{in})}, \quad (16)$$

where p_{ex}, p_{in} are the mass-weighted static pressure at the diffuser exit and reference location (mixing tube location), ρ_{in}, U_{in} are density and axial velocity at mixing tube inlet.

Figure 11 shows the value of overall C_p as a function of swirl number. It can be seen that the presence of any amount of swirl in the flow tends to improve the value of overall C_p compared to no swirl case S=0. Hence, it is concluded that overall C_p value increases for an inline-slot ejector diffuser for moderate swirl flows.

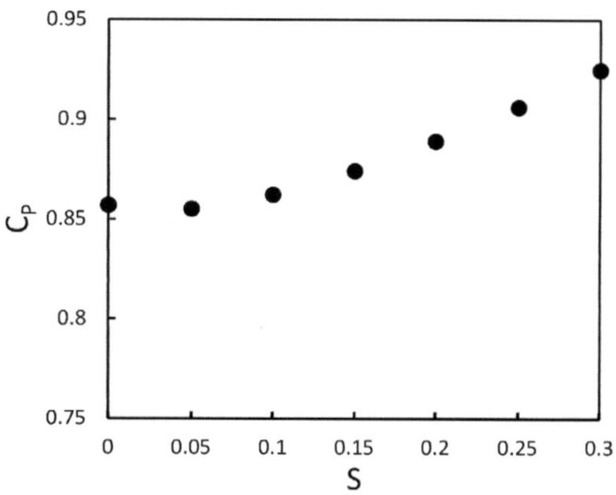

Figure 11: Static pressure recovery coefficient for varying nozzle inlet swirl number.

4.2 EFFECT OF INLET TURBULENCE INTENSITY

The turbulence intensity (TI) at the nozzle inlet is governed by the gas turbine exit conditions and is given by Equation 17

$$TI = \frac{u'}{U} = \sqrt{\frac{2k}{3}} \qquad (17)$$

where k represents the mean turbulent kinetic energy. In the present investigation TI is varied in the range $1\% \leq TI \leq 15\%$ in steps of 3 to study its effect on the performance of the inline-slot ejector diffuser.

4.2 (a) Mass Entrainment

The definition of local and cumulative mass entrainment ratio is explained in Section 4.1 (a). Table 5 displays the variation in κ with change in TI for all the slots. The effect of turbulent intensity on κ shows that (i) mass entrainment through the standoff distance increases as nozzle inlet TI increases. An increase of ~2% in κ is observed when TI is 15% compared to 1%. (ii) no clear trend of κ is observed at first slot. (iii) From second slot onwards, the value of κ decreases with increase in turbulent intensity. At the fifth slot κ for the TI =15% drops by ~5.5% compared to the case TI=1%.

Table 5: Comparison of local mass entrainment ratio (κ)

	TI_1	TI_3	TI_6	TI_9	TI_12	TI_15
κ_{SD}	0.9784	0.9805	0.984	0.9884	0.9921	0.9951
κ_1	0.1522	0.1523	0.1523	0.152	0.1515	0.1509
κ_2	0.2946	0.2934	0.2914	0.2884	0.2863	0.2844
κ_3	0.3908	0.3886	0.3844	0.3799	0.3763	0.3733
κ_4	0.5223	0.5199	0.5146	0.5095	0.5044	0.4996
κ_5	0.8992	0.8767	0.8697	0.8632	0.8561	0.85

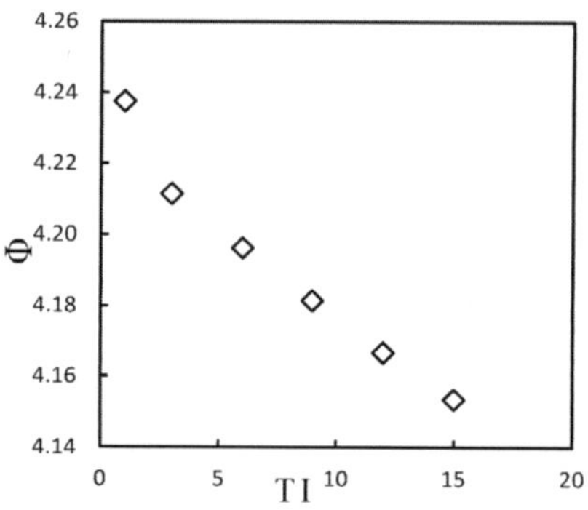

Figure 12: Cumulative mass entrainment ratio (φ) variation

Figure 12 shows the cumulative mass entrainment ratio (φ) as a function of TI and it can be seen that with the increase of turbulence intensity from 1% to 15% the value of φ decreases. At the end of the fifth slot, φ=4.153 for TI=15% has minimum while φ=4.238 for TI=1% is maximum. Thus, it can be concluded that cumulative mass entrainment decreases with increase in turbulence intensity.

4.2 (b) Temperature Distribution

Analysis of the temperature distribution parameter (ψ) reveals no significant changes when turbulent intensity is varied from 1% to 15%. Figure 13 shows that the profile of ψ for all cases overlaps with each other.

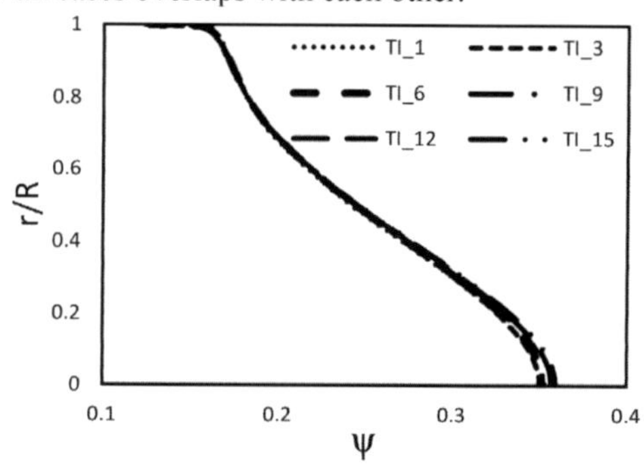

Figure 13: ψ at the exit of the ejector diffuser.

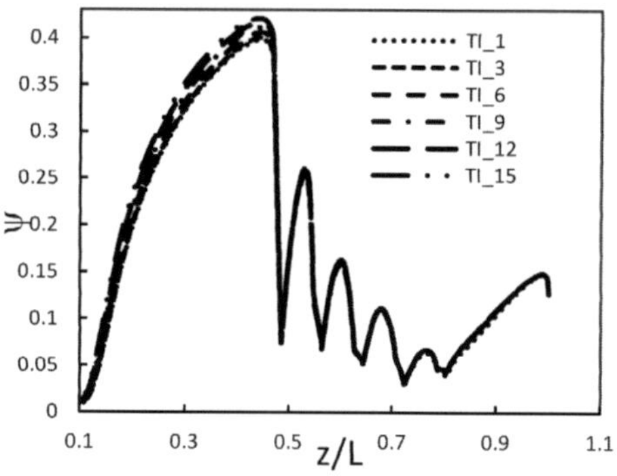

Figure 14: Wall temperature distribution under varying turbulent intensity

Figure 14 shows the wall temperature in terms of ψ under varying turbulence intensity cases. Analysis of ψ at the wall shows that for $z/L < 0.44$ which corresponds to the mixing wall, slight variation in the value of ψ is seen where slightly higher temperature is obtained for higher turbulence intensity case. This could be due to higher mixing for turbulent flow. However, no such influence of turbulence is seen at the walls of the diffuser.

4.2(c) Static Pressure Recovery

Figure 15 shows the variation of C_p as a function of turbulence intensity. It can be seen that no significant variation in C_p is observed when turbulence intensity is changed from 1% to 15%. So, it can be concluded that the static pressure recovery is not a function of turbulence intensity.

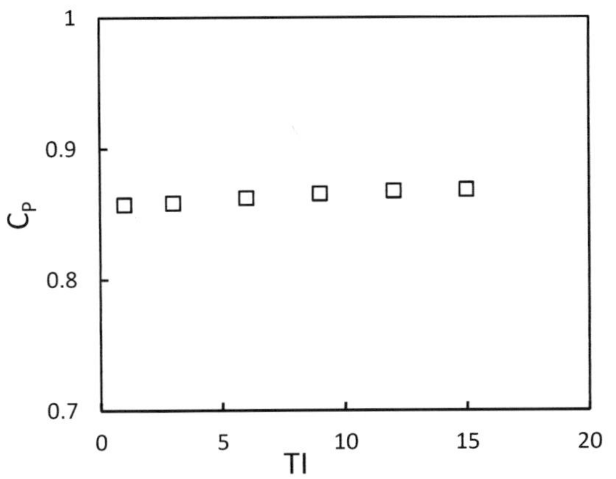

Figure 15: Static pressure recovery for varying nozzle inlet turbulence intensity.

7. CONCLUSIONS

The current study has been conducted to understand the effect of nozzle inlet conditions (swirl and turbulence intensity) on the performance of an inline-slot ejector diffuser. The work is a carried out by numerically solving the governing equations using the software Ansys Fluent. The study has been conducted at the fixed nozzle exit Reynolds number of 5×10^4. The nozzle exit temperature is fixed at 700K while the ambient temperature is fixed at 300K. All air properties are defined as function of temperature. In the first part of the study, swirl number is varied from 0 to 0.3 in the step of 0.05. This represent flow condition of no swirl to moderate swirl. In the second part, turbulence intensity is varied from 1% to 15% in steps of 3%. This represent the flow conditions at the nozzle exit from low turbulence level to extreme turbulence cases. It is seen that with the increase in swirl number from 0 to 0.3, 3% drop in cumulative mass entrainment is observed. As swirl number increases the potential core region is reduced. Higher mixing tube wall temperatures are observed for all the configurations with swirl cases. Static pressure recovery increases with the increase in the swirl number. Analysis of the increase in turbulence intensity shows that the mass entrainment decreases by ~5% when turbulence intensity is increased from 1% to 15%. No significant effect of turbulence intensity is seen on the temperature distribution and pressure recovery.

9. REFERENCES

1. Birk AM and VanDam D. Marine gas turbine infrared signature suppression: Aerothermal design considerations. *International Gas Turbine and Aeroengine Congress and Exposition*, pages V002T03A009–V002T03A009–10, 1989.
2. Rao AG. Infrared Signature Modeling and Analysis of Aircraft Plume. *International Journal of Turbo and Jet Engines*, 28:187, 2011.
3. Sen S. Studies on Flow Characteristics of a Stepped Conical Diffuser with Passive Suction. PhD thesis, *Department of Applied Mechanics, Indian Institute of Technology Delhi*, 2008.
4. Singh P, Singh S N, and Seshadri V. Experimental Investigations on Non-Circular Ejector Air Diffusers. *39th AIAA Fluid Dynamics Conference*, page 4213, 2009.
5. Singh P, Singh S N, and Seshadri V. Effect of number of slots and overlap on the performance of non-circular ejector air diffuser. In *43rd AIAA Fluid Dynamics Conference*, page 2729, 2013.
6. Chen Q. Performance of air-air ejectors with multi-ring entraining diffusers. PhD thesis, *Department of Mechanical and Materials Engineering, Queen's University Canada*, 2008.
7. Singh, L., Singh, SN., and Sinha, SS., 'Effect of slot-guidance and slot-area on air entrainment in a conical ejector diffuser for infrared suppression' *Journal of Applied Fluid Mechanics*, 12(4):1303–1318, 2019.
8. Singh, L., Singh, SN., and Sinha, SS., 'Effect of standoff distance and area ratio on the performance of circular exhaust ejector using computational fluid dynamics' *Proc. IMechE, Part G: Journal of Aerospace Engineering*, 232(15):2821–2832, 2018.
9. Maqsood, A., 'A study of subsonic air-air ejector with short bent mixing tube. PhD thesis, *Department of Mechanical and Materials Engineering', Queen's University at Kingston, Canada*, 2008.
10. Menter, FR., 'Two-equation eddy- viscosity turbulence models for engineering applications', *Journal of AIAA*, 32(8):1598-1605, 1994.
11. Simone Crippa. Improvement of unstructured computational fluid dynamics simulations through novel mesh generation methodologies. Journal of Aircraft, 48(3):1036–1044, 2011.
12. W. B. Nicoll and B. R. Ramaprian. Performance of conical diffusers with annular injection at inlet. Journal of Fluids Engineering, 92(4):827–835, 1970.

9. AUTHORS BIOGRAPHY

Lakhvinder Singh holds the current position of Research Scholar at Indian Institute of Technology Delhi. He is a final year PhD student. He works in the area of passive infrared suppressors used in combat vehicles such as helicopters, warships, tanks, etc.

S. N. Singh holds the current position of Professor in the Department of Applied Mechanics, Indian Institute of Technology Delhi. He works in the area of internal Flows: diffusers, combustors & pipe flow; Computational Fluid Dynamics of Internal Flows; Flow Instrumentation; Open Channel Flow.

S. S. Sinha holds the current position of Associate Professor in the Department of Applied Mechanics, Indian Institute of Technology Delhi. He works in the area of Theory and Modelling of Compressible Turbulence, Hypersonic Flows, Airships, Flow-thermodynamics of steam turbines, Scale-resolving methods of turbulence computations

AUTHOR INDEX

The Royal Institution of Naval Architects
8-9 Northumberland Street
London, WC2N 5DA
United Kingdom

ISBN 978-1-5108-9542-3

CURRAN ASSOCIATES INC.
proceedings.com